Patterns of Human Variation

Patterns of Human Variation

The Demography, Genetics, and Phenetics of Bougainville Islanders

Jonathan Scott Friedlaender

Harvard University Press   Cambridge, Massachusetts, and London, England   1975

To Bilgé

Foreword

Population genetics is widely regarded as the best example of the power of theoretical formulations in biology. There is no doubt that the theory of population genetics is highly developed, and even hypertrophied. Given the rates of gene mutation, pattern of migration between populations, system of mating within populations, population sizes, intensities of natural selection, then it is possible to predict changes in gene frequency. If this prediction cannot be made by the solutions of differential equations, then at least numerical and simulation studies can make specific numerical predictions. But despite this highly developed theory, there has been very little successful specific explanation of particular polymorphisms in particular species. The reason is that it is exceedingly difficult to measure the mutation rates, migration patterns, reproductive structure, population size and components of reproductive success in any given species. For some species, we know a lot about formal genetics (Drosophila, for example), but little or nothing about its demography or migration. For others, we know rather more about their demography (univoltine insects, for example), but it requires very special conditions to determine the migration between populations or the population sizes. It is not always realized by population geneticists whose training has been with experimental animals and plants, or with the more usual objects of natural historical study, that there is one species about which there is vastly more information than any other. That species is man. We have extraordinarily detailed and accurate information on the age-specific mortality and fecundity rates, the morbidity patterns, the migration patterns, mating behaviour, general life habits, and *history* of the human species. Even for remote and poorly known indigenous populations of South America and the Pacific and Indian Oceans, we can learn, and indeed already know, incomparably more than for any other species, no matter how well studied. Thus, it is in man, above all other species, that we are likely to answer some of the leading questions of evolutionary and population genetics.

Professor Friedlaender's book is a fine example of what can be done in human populations to understand the forces operating on genetic variation. Only one other study, that of the Xavante and Yanamamö Indians by the University of Michigan group, is as detailed and all-encompassing. What Professor Friedlaender tries to do is to establish the actual degree of relationship among villages of a group of related peoples by reference to their language, geographical position, and patterns of movement, and then to ask whether the genetic differentiation among these villages reflects that history of relationship. If there is a poor fit between the pattern of historical relations and the pattern of genetic similarities, then we must assume that natural selection, operating differently in different localities, has contributed significantly to the observed genetic variation. If, on the other hand, the fit is good, we can assume that only the forces of migration and random population differentiation have been involved. Thus, the simultaneous study of the history and genetics of a group of human populations can help to answer the leading question of evolutionary genetics today: What are the relative roles of natural selection and random processes in modling the genetic variation in species?

Professor Friedlaender's work shows both the virtues of studying human populations and the defects. Perhaps chief among the defects is the great weight that anthropologists must put on a rather too simple measure of linguistic similarity, the percentage of cognates in two languages. It is a pecularity of the entire method of investigation that one set of characters, the genetic ones, with an extremely sophisticated theory of their evolution, is being compared with another set, linguistic, with an extremely crude theory of evolution. One possible side effect of such studies as this one would be an impetus to develop a more sophisticated quantitative theory of linguistic evolution and divergence.

The great virtue of the study of human populations is that it introduces *history* into evolution. This seems a strange assertion since what is evolution if not the history of life. Is not the study of evolution always necessarily historical? Curiously, the answer is no, not usually. Most modern theoretical, experimental and naturalistic studies of evolutionary phenomena are specifically anti-historical because they are mostly concerned with *equilibrium* populations and equilibria destroy history. Even when populations are not known to be in equilibrium, it is assumed that they are, because it is impossible to reconstruct history for populations of fruit flies. For butterflies, a little history can be gleaned from museum collections and for snails rather more from deposits of shells. But, in general, history of wild animals and plants is hard to come by. In the study of man, however, we make the opposite assumption. We are keenly aware of the historical contingency of each human situation (at least that is true for biologists and anthropologists. Other social scientists, especially in

Western countries, have an absurdly ahistorical view of psychology, sociology, and political economy). The historical perspective on an evolutionary phenomenon permeates Professor Friedlaender's book as it must any rational study of human biology. While there are severe limits to the historical information that can be obtained or inferred for these nonliterate peoples, it is not much different for those with a rich written record. For all our histories and chronicles we know precious little about the genetically effective migrations and demographic changes in Europe since Roman times. Yet, despite the imperfections of the information, studies like Professor Friedlaender's are absolutely essential, and are extremely illuminating both for questions of general evolutionary interest and for a knowledge of our own species.

R.C. Lewontin

It has been my pleasure and delight over the last seven years to become immersed in this study of human variation among Bougainville Islanders. It all began quite inauspiciously. I had completed my Ph.D. qualifying examinations at Harvard and was looking rather desperately around for a thesis project, when Douglas Oliver, whose association with Bougainville goes back to 1938, casually suggested that I tag along on the Harvard Solomon Islands Expedition in the summer of 1966. I had tried unsuccessfully to go on that trip, but now that I had a National Science Foundation Fellowship with its own financing, I suddenly became an asset to the team of ten medical men and cultural anthropologists. Fellowships and grants are like olives in a bottle. After the first one, the rest is easy.

During that two-month stay in Rumba, Bougainville and Sinerango, Malaita, I learned most of what I needed to know in order to carry out my own field survey. Although I borrowed or learned something from almost everyone on the team, I must single out the most important. Albert Damon, the indefatigable expedition leader who was even then suffering from a malignancy which took his life last year, was most generous with the expedition equipment and data, and constantly sought to cut through my hazy notions with sharp and pertinent questioning. Eugene Ogan introduced me to the people and societies of Bougainville and gave me a great deal of sound advice as to how to conduct myself in that land. Eugene Giles, who had carried out a similar survey in New Guinea, helped me decide what measurements, tests, and numbers of people could be covered in how much time. Jerome Bloom taught me the rudiments of taking blood samples.

During the summer expedition, I applied for a NSF Dissertation Improvement Award which would enable me to undertake an independent survey of what I originally estimated to be an additional 700 individuals in the area immediately to the north of Arawa, Bougainville, where members of three very different language groups (Eivo, Nasioi, and Torau) lived in close proximity. I received word in September that I could proceed, and spent the next seven months conducting the survey which forms the basis of this book, covering seventeen villages and something over 2,000 individuals.

I rapidly developed a standard procedure for my work. In areas which had decent parish birth and death records (the Eivo, Simeku, Nasioi, and Torau),

I collected family and village census information from the mission before beginning work in the village. These formed the basis for many of the inferences on the demography of the area in Chapter 3, and helped me to locate and to assign ages to people once in the village. Ordinarily I would proceed to the village with my 300 pounds of equipment by foot and set up "camp" in a house in the village. Anywhere from ten to thirty people would help me carry my equipment, depending not so much on how much there was to carry as on how many people wanted to earn thirty cents an hour carrying it.

Upon my arrival in a village I generally was asked to explain what I was trying to do and what I wanted of the people. I attempted as well as I could to explain that I was interested in the physical differences between the various groups on Bougainville and how they related to the language relationships and patterns of intermarriage of the people. Also, that blood and fingerprints were aspects of the physique which reflected hereditary relationships among groups, and this was why I was interested in them. I also explained that the sort of linguistic diversity and marital exchange patterns which characterized them were totally different from those in America and almost all other countries in the world, which was what made them interesting to me. I also explained that, while their blood and measurements would not make me rich, they would enable me to write a book about all of these things which I hoped would finally get me out of school and get me a good job. Beyond all this, I could not offer anyone payment for cooperation, but could only give families a polaroid photograph as a gift in return for their help. (The medical service was afraid that if I paid subjects, that payment would be demanded for future medical surveys.)

Remarkably enough, only one group I visited (Vito, a Torau-speaking village) decided that they did not want to be subjected to my little indignities. Except for some of the most isolated villages, where the whole episode must have seemed incomprehensible, I am convinced most people understood the basic aims of my work. A number of people acknowledged that they were quite aware of the physical differences of different groups on the island and had wondered themselves how they had come about. A couple of elder leaders told me their own ideas on the nature of the differences in physique of the different areas of Bougainville, which correspond quite well with the final picture that I derived from the multiple discriminant analysis of the anthropometric measurements. The educated eye is not to be denied.

Ordinarily, I began interviewing families the day following my arrival. The specifics of the data-gathering battery appear in some detail in the Appendix, but it should suffice here to say that, with the sort of maximum cooperation I received, I could conduct a cursory demographic interview, take measurements on adult males, take fingerprints on all residents, and carry out tests for PTC taste sensitivity, color blindness, and visual acuity on approximately fifty people a day. After all the available residents of a village had been covered, and a genealogy of the village constructed, I took 10 cc blood samples from everyone old enough to have palpable antecubital veins (2006 individuals). No one likes to give blood, and the people of Bougainville are no exception. Nevertheless, I got maximum cooperation from them in this regard once again. It was, of course, partially due to the expectation of receiving a polaroid photograph of themselves, but also because of the novelty and sense of self-importance which the entire survey seemed to give the villagers. Everyone remembered their particular experiences with the survey vividly, so that when I visited some villages again in 1970 and 1973, people were always reminding me of things I did or particular incidents that I had completely forgotten. When I asked a man in 1970 the name of his father, his response was, "Don't you remember? You asked me that before."

Of course, my work would have been impossible without the help of my assistants, who enabled me to gain access to many villages, sometimes recorded anthropometric measurements, and helped in the general organization and data-gathering process. They are Hilary Arlaw, Francis Dipawe, Ignatius Bakamori, Michael Madaku, and Maneha.

Colin Murray, Shirley MacDonald, and Roger Porteous saw to it that the blood samples got special treatment on their way to Sydney.

Within the Bougainville District bureaucracy and Public Health Department, Max Denehy, Philip Hardy, William Brown, and Dr. Lindsey were all of assistance.

The Marist Fathers were a great source of support, both logistically and psychologically, throughout my stay. I came to depend upon Fathers Nicholas Kutulas, Robert Wiley, and Robert Pelletier for initial contact with the people in their respective parishes, and once there, I depended upon the mission for contact with the outside world. Without their *Stati Animarae,* I would have no decent age records for this study and no check on my own census figures. Sister

Marie Augusta, Father Lebel, and Father Miltrup were also very helpful. I would also like to acknowledge here (more explicitly in the Appendix) the irreplaceable role of Professor R. J. Walsh and his laboratory at the Australian Red Cross and later at the University of New South Wales. I shipped all blood samples to them, and they carried out various analyses in remarkably good order and good humor. Dr. Arthur G. Steinberg, of the Case Western Reserve University, Cleveland, carried out the analyses of Gm and Inv factors in fast and efficient style.

In 1970, I was able to return to Bougainville, along with Dr. Howard Bailit (a dental researcher on the 1966 Harvard Solomons Project) and his student, Glenn Rapoport. During this shorter two-month stay, I was able to collect more detailed genealogies on the northeast Siwai while Bailit and Rapoport conducted an extremely fast and efficient survey (including taking dental impressions and making plaster casts) through the east coast villages I had covered in 1966–67. It is this later survey which forms the substance of Chapter 7, which was also partially written by Howard Bailit. Although he deserves co-authorship of this book, he has most graciously declined.

Dr. Jeffrey Froehlich supervised the analysis of the finger and palm prints, carried out by Brian Eisenberg and Christopher Hitt. I have relied heavily on his expertise and judgment in Chapter 6.

Over the last six years I have subjected the large body of data to a wide variety of different statistical analyses which seemed suitable for the questions I was interested in. This has always meant relying on the knowledge and assistance of others, especially in the early stages of any analysis. Kenneth Jones, now of Brandeis University, introduced me to multivariate statistics. Henry Harpending, now of New Mexico, has been a major source of fruitful discussions and ideas over this time, beginning with the week I returned from the field. Newton Morton, of the University of Hawaii, has had a strong influence on my thinking, as my recent papers attest (Friedlaender, 1971a, 1971b, and 1974). Laura Sgaramella-Zonta and Kenneth Kidd analyzed the data from a different perspective, which has also been productive (see esp. Friedlaender et al., 1971). Finally, John Rhoads, currently a graduate student at Harvard, has shepherded the data through the final stages of "perfection" visible here.

It goes without saying that the concept of this entire study derives from previous work of my graduate advisor and now colleague, William Howells. His own interest in Melanesia goes back to his doctoral dissertation, and more recent papers of his served as the framework for this project. Richard Osborne pointed out the possibilities of the restudy of men previously measured by Douglas Oliver. Others whose advice I have also sought include James Crow, Charles Cotterman, Sewall Wright, William Bossert, John Grove, Marcus Feldman, and Robert Elston.

A number of people have read various drafts of this book, and while I have not always agreed with their ideas, I would like to thank those who made specific suggestions: Douglas Oliver (Chapter 2), Eugene Ogan (Chapter 2), John Terrell (Chapter 2), Donald Mitchell (Chapter 2), Stephen Gould (various), Diane Wagener (Chapter 3), Luca Cavalli-Sforza (Chapter 3), and Richard Lewontin (Chapter 4).

Among those who have helped in various ways in the preparation of the data and manuscript, I must mention Mary Hyde, Judith Selvidge, Judith Neal, Elizabeth Gude, Bonibel Sack, Donald Mitchell, Virginia Hildreth, Susan Davis, Catherine Kazar, Nancy Lubin, Kathleen McGrath, Buffy Ellis, Susan Groisser, and Harriet Spiegel.

I should explicitly mention the support I have received from the National Science Foundation, both in the form of a graduate fellowship, a generous dissertation improvement award, and a post-doctoral grant (GS-3088), the Population Council Post-Doctoral Fellowship I received at the University of Wisconsin, indirect support from NIGMS 13482, and Harvard grants from the Teschemacher, Milton, and Graduate Society Funds.

J.S.F.

Acknowledgments

I would like to express special thanks and appreciation to my colleagues, who carried out with great efficiency phases of the data collection and analysis:

Howard L. Bailit

Robert J. Walsh

Arthur G. Steinberg

John G. Rhoads

Jeffrey Froehlich

Contents

Figures

Plates

27–62. Twice Measured Males (photographs taken in 1938–39 by D. L. Oliver and in 1967 by J. S. Friedlaender)

Tables

Patterns of Human Variation

1 Introduction

How does one group of people come to be different from another? If groups are different with respect to one sort of physical character (for instance, blood types), are they likely to show similar differences in other aspects of the physique? Can one predict the pattern or degree of physical differences among groups if the current pattern of intermarriage among them is known? And do linguistic or historical relationships among groups closely reflect the biological ones? These are some of the questions I will attempt to supply rudimentary answers for, at least for one small but incredibly diverse set of villages on Bougainville Island, in the new nation of Papua New Guinea.

The questions are not new; patterns of human zoogeography have been a major preoccupation of physical anthropologists for a number of generations. But their context and their interpretation have changed markedly during the last thirty years. The early framework for such studies was racial classification and taxonomy. Anthropologists took their task to be the elucidation of all the varieties of Man, with the implicit assumption that a definitive racial history would emerge in time. Essentially since the Second World War, anthropologists have adopted some sort of evolutionary context for their studies. But as they were a little late in adopting an already existing body of theory and established principles from evolutionary biologists, there was the sense that they were constantly trying to catch up with their more sophisticated scientific brethren. Interestingly enough, as the focus of evolutionary theory has begun to shift in the past decade, the unique features of human population variation have placed such studies, and the old questions, on center stage again.

The Dilemma of Modern Evolutionary Theory

It is conventional to divide the evolutionary process into two steps; the production of variation among individuals within a population (through mutation, recombination of existing genes, migration of foreign individuals, and random sampling events), and the action upon that variation by natural selection. That is, not all variants are equally equipped to live and reproduce successfully in a given situation, and those less well endowed will, on the average, leave fewer and fewer descendants.

Beginning with Darwin, who introduced the theory of natural selection as the guiding principle in evolution, there has been some difficulty in explaining or accounting for the origin of new variation in different characters in different populations or even different species. During the first decade of this century, some experimental geneticists (particularly deVries) became so impressed with what appeared to be variation caused by spontaneous mutation that they discounted the ordering role of natural selection in evolution, arguing that new species arose through gross mutations. It was during this period that the division between the experimental geneticists and naturalists, the mutationists and selectionists, was clearest. Some scientists, such as Bateson, suggested there might be different genetic mechanisms controlling different sorts of characters in the phenotype; the discrete, qualitative traits that preoccupied experimental geneticists being controlled by Mendelian segregating genes, and the more generally obvious, normally distributed traits, which naturalists saw as clear products of some organism-environment interaction, being the results of mechanisms of blending inheritance and the action of natural selection.

This conflict was, to a large degree, resolved in the following decades through the work of a number of brilliant men, but perhaps most notably by R. A. Fisher. He showed that variation in normally distributed traits, such as stature, could be determined by the action of a number of separate genes. No recourse to blending inheritance was necessary. Also, natural selection, in acting to eliminate organisms at one end of such a continuous distribution, could lead to gradual genetic change over time. The laboratory geneticists accepted selection, and the naturalists embraced particulate genes.

The new synthesis, or neo-Darwinism, has been virtually unchallenged dogma through most of the last thirty years. By and large, the stress has been on the primacy of selection in molding evolutionary change. Mutation, migration, recombination, and random sampling events (genetic drift) become largely predictable phenomena, while selection is the active agent, the true shaping force. George Gaylord Simpson, for example, has stated that selection is the *composer* of organic variation, implying that mutation and the rest are available keys on the piano or notes in the air to be utilized at will.

In the last few years, however, there has been a good deal of questioning of the sufficiency of the strict selectionist position. On the simplest level, there appears to be a fairly good chance that large sections of the genetic material may serve as nothing more than spacing material between active gene sites, and as such would be the object of only the most meager form of natural selection (Kohne 1970). Another enticing development has been the establishment of the structures of various proteins in different organisms. It seems that, at least for some sections of proteins, the alterations in structure among different species have followed quite uniform rates of "decay," or rates of divergence from each other, with the most dissimilar structures belonging to the most distantly related species (e.g., Fitch and Margoliash 1967). Some have argued that this is not what one would expect if such a protean force as natural selection were the overriding determinant. Even more radical is the suggestion that the inherent redundancy of elements of the DNA code determines the relative frequency of particular amino acid residues in proteins; those amino acids with the largest number of different codons appear most commonly (King and Jukes 1969). Again, here is the idea that selection may not be responsible for a number of evolutionary phenomena—that many sorts of variation may not be open to the action of selection.

Genetic Variation within Local Populations

If the overriding importance of selection has been questioned anew at the level of the codon, the chromosome, and amino acid residue, it has been attacked as the major limiting force of variation at the level of blood polymorphisms.

There is an absolutely enormous amount of genetic heterogeneity, or heterozygosity, in most human populations (and most other organisms as well).[1] Judging from enzyme electrophoretic studies, it would appear that at least 30 percent of the active structural genetic loci are variable, or polymorphic. While naturalists recognized the hidden reservoirs of genetic variability in populations, for many population geneticists evolution in its most reductionist sense was equated with the replacement and fixation of one gene at the expense of another. Selection was assumed most often to favor one gene over the alternatives in a particular population and ecological situation, leading to the preponderance of that gene. Selection can, of course, favor combinations of genes in the same

1. Harris and Hopkinson (1972) have found that 21 of 70 loci in the "English population" are electrophoretically distinguishable as polymorphic but three times that many may be variable and be undetectable by this method.

individual or in the same population (for instance, the classic example of sickle cell trait in a malarial environment), but it has been hard to imagine so much genetic diversity being maintained by such a mechanism, where in each generation so many individuals who do not have the proper combination will be at a relative disadvantage. However, recently it has been shown (Johnson 1974, Prakash 1973) that classes of enzymes which participate in chemical reactions involving substances coming directly from the environment tend to be much more variable in their structure than those participating in more "internal" metabolic processes. This suggests that increased heterozygosity and protein variability may be a selective response to environmental variability after all. There are other possible explanations: selection may be operating to maintain different genes in different populations; we may often be sampling a transitional state, where one gene is in fact being replaced by another; combinations of genes at different sites may be selectively superior, and so on. At the moment, very little is certain about these important matters. Except for very obvious cases of selection (as with sickle-cell hemoglobin) and mutation and drift (as in elevated frequencies of various genetically determined abnormalities in isolates such as the Amish), most genetic variation in restricted populations can be interpreted equally satisfactorily to suit the most strict selectionist or ardent neutralist.

The problem is compounded because the relative effects upon gene frequencies of weak selection, population size, and mutation may be indistinguishable. As Sewall Wright said long ago (Wright 1948), only when a directional selection coefficient or mutation pressure is greater than the proportion $1/4N$ (with N equaling the effective size of the breeding population), that force will predominate. But if those pressures are less, or when $1/4N$ is large, as in very small isolates, then sampling effects will predominate.

One recurring pattern of variation in animal species which is explicable by this argument is the phenomenon of high genetic variability in the center of a species range and the relatively low variability on the periphery (Mayr 1963). Rather than invoking selectionist arguments concerning the relative breadth of adaptation of organisms to differing ecologies within the species range (Van Valen 1965, Slatkin 1970, Mayr 1963), we must consider the possibility that the reduced amount of migration among the more isolated groups on the periphery accounts for their relative uniformity. Conversely, the greater rates of migration within the central region of a species range (and from the periphery) can itself account for the greater genetic heterogeneity in that area (see Rohlf and Schnell [1970] for a computer simulation of Wright's 1969 model of isolation by distance).

When detailed attempts are made to account for particular patterns of genetic diversity by invoking the dynamics of migration, historical accident, population size, age distributions, and distributions of progeny size, then human populations, and particularly small, relatively isolated human populations, become especially informative. In recent years, physical anthropologists and human geneticists have accumulated an impressive body of data on blood-group and serum protein variation in arrays of small, semi-isolated communities in a wide variety of populations. Besides earlier reports on genetic differentiation in Bougainville, excellent surveys of blood polymorphism and migrational variation have been carried out in South and North American Indian populations, mestizo groups, Australian Aborigines, Micronesians, African pygmies, and African Bushmen.[2] In all of these studies, and in many other such blood polymorphism surveys, a truly remarkable degree of genetic heterogeneity within and among localized groups manifests itself. When variation at a number of different gene loci are considered together, the pattern of variation closely approximates a pattern or degree of variation predicted from what is known of migration and historical relations among the local groups. Furthermore, it appears that the degree of variation is largely predictable from the nature of migration within the total populations, with the most diverse variation found in small, stable groups tied to the land, most notably in tropical gardeners (for example, Bougainville and New Guinea Highlanders), less in more mobile hunter-gatherer bands (African Bushmen, Pygmies, Greenland Eskimos), and by far the least in cosmopolitan areas where communication and migration are most developed. Cavalli-Sforza said recently (1973) that in microgeographic studies, in the controversy over whether drift (including migration and all associated demographic quantities) or selection is the causative factor in variation, drift wins. Genetic

2. The references are, respectively, Friedlaender (1971a, b), Neel and Ward (1972), Workman and Niswander (1970), Morton et al. (1966), Kirk and Blake (1971), Morton et al. (1972), Cavalli-Sforza et al. (1969), Harpending and Jenkins (1973).

variation from one continent to the next is a different matter, and almost no one accounts for all of these patterns without some recourse to selection (see, for example, Cavalli-Sforza 1966, and Lewotin and Krakauer 1973).

In interpreting the patterns of variation among these Bougainville Islanders, I will focus on possible nonselective determinants of the patterns. This hardly should be interpreted as indicating that I am a "neutralist," but rather that this is the simplest and most logical path to follow. I know *something* about rates of migration in the area, population sizes, and historical relationships of the groups, but almost *nothing* about the nature of differential selection pressures which might account for the great genetic diversity within the island's population. It is difficult for me to conceive of a selectionist explanation for all this diversity as well. If the biological variation generally "fits" the pattern I predict, it hardly proves that selection is not operative or instrumental in causing the pattern of diversity. To make such a judgment demands an alternative model which I am totally incapable of formulating.

Phenotypic Variation within Local Populations

Although a good deal of progress has been made in elucidating the blood genetic diversity of small populations, aspects of the human phenotype which are under more complex genetic and environmental control have been relatively neglected in the context of this problem. And no wonder. The relationship of genetic variation and heterozygosity with more complex phenotypic variation has yet to be decently understood for any complex organism.

On one side there is the belief that heterozygosity, and therefore genetic variation within a population, acts to reduce some elements of phenotypic variation by providing alternative developmental pathways for different systems, or improving developmental homeostasis (Lerner 1954). Also, intense selection may lead to increased variability in a wide number of characters. However, this general approach is based upon experiments with highly inbred, homozygous animals which are quite unlike any human populations with regard to the levels of inbreeding, or to their artificially instituted regimen of selection.

On the other side, Fisher and many other geneticists have assumed that phenotypic variation is directly proportional to the amount of genetic variation and heterozygosity in a particular population. Fisher also thought that where there was very little variation within a population, there was evidence for strong selection; and, conversely, where there were high levels of variation, there was presumptive evidence for less intense selection. In sum, our present ignorance of developmental genetics makes it very difficult, if not impossible, to interpret degrees of variability of characters as being the result of more or less selection, or more or less genetic variability.

Soulé (1971) has explained the pattern of phenotypic heterogeneity in lizards living on islands off the Baja California coast by invoking the same non-selective agents now commonly invoked in the literature of human population genetics—drift and migration. He found a good correspondence between the amount of phenotypic variation (as manifested in scale counts at various points) and genetic variation in enzyme polymorphisms.

In the anthropological literature there have been some implicit associations of variation in genetic characters with phenotypic variation. Pollitzer made genetic maps and anthropometric maps of selected American white, black, and African black populations, getting similar pictures. Littlewood (1973) has obtained crude agreement in anthropometric and blood group variation in the New Guinea Eastern Highlands.

The Survey Model and Hypothesis

The survey I conducted through central sections of Bougainville Island has special significance for these various problems because of the intensive nature of the sampling, the extraordinary diversity of the populations covered in a small section of the island, and the wide variety of biological variables included in the study.

The central section of Bougainville, particularly on the east coast, has as diverse a group of indigenous languages as any comparable area of which I am aware (see Figure 2.3). In this area, there are representatives of three very different language stocks, the northern and southern Bougainville Papuan stocks of the interior, related at a very low level, and the coastal representatives of the Melanesian division of the Austronesian languages.

These language clusters were the primary focus of my interest, and I set out in September 1966 to cover a minimum of two "villages" from each language area (providing for some estimation of both within- and among-language group variation), but I finally surveyed more villages in order to complete "linkages" among the villages. As pictured in Figure 2.3, villages 3–13 form a tight group and can be said to occupy adjacent positions along a single path (refer also to Figure 2.2). Villages 17 and 18, on the beach, have varying degrees of contact with each other and their inland neighbors, and they have had significant contacts with the southern offshore islands, in some cases as recently as 100 years ago. These thirteen villages provided the information necessary for the reconstruction of marriage exchanges and short-range migrations which would serve as predictors of biological relationships among the groups.

To this basic cluster I added two villages from the adjacent northern Papuan language group (Rotokas and its Aita dialect), just to the north of an uninhabited stretch of land. I hoped these would serve as a polar reference for the biological variation of the more southerly groups on the east coast, all of which were clearly exchanging significant numbers of mates. Finally, I included villages from the Siwai area to the southwest, primarily because they offered a special opportunity for a longitudinal study of aging effects and the secular trend (see Chapter 5), and because they might present an analogous southern "pole" for the variation. Moreover, while the Siwai are removed from the east coast groups both in terms of measurable migration and in terms of geography, linguistically they are closely related to the Nasioi, represented by samples 11–13. This offered an interesting variation which might help to distinguish the more important correlates of biological diversity in the area.

As I envisioned the study, I hoped to define the patterns of current migration and past relationships (largely deduced by language affiliations) in an early stage of analysis (see Chapter 3). Then, some measure of genetic variation, as reflected in blood polymorphism frequencies, could be established over the survey populations and related to the demographic findings (Chapter 4). Finally, it should be possible to compare these various patterns with those established with data from other genetically controlled or influenced aspects of the phenotype: (1) anthropometrics, features which are relatively poorly genetically controlled, open as most of them are to environmental influences during the life span, to aging, and to selection pressures of all kinds as well (Chapter 5); (2) the finger and palm ridge patterns, supposedly under extremely close control by a number of genes and subject to little direct selection pressure

(Chapter 6); and (3) tooth size, presumably open to more direct selection than dermatoglyphics, but still under fairly close genetic determination and control (Chapter 7).

Sampling procedures in the villages were necessarily different for each data set, but completeness and maximum overlap from one set to the next were the goals. Table 1.1 gives the different sample sizes, the census sizes of the villages three years prior to the first survey, and the dates of the surveys for each village.

The actual methods for data-gathering and analysis are described in the Appendix and in the appropriate chapters.

The various data sets (derived from simple geographic distances, migration information, language relationships, anthropometrics, male and female den-

Table 1.1
Census and Sample Sizes of the Bougainville Village Series

Village	1966–67 Survey beginning date	(1963) Census size	Red blood cell sample	Dermatoglyphics Male	Dermatoglyphics Female	(1970) Dentometrics Male	(1970) Dentometrics Female	Anthropometrics Male
Nupatoro	3/7/67	164	116	22	19	15	9	21
Okowopaia	1/24/67	128	98	24	23	20	18	19
Kopani	9/12/66	245	180	59	79	43	56	54
Kopikiri	9/26/66	95	53	22	22	22	26	10
Nasiwoiwa	9/30/66	135	101	45	42	41	47	36
Atamo	10/18/66 ⎫	242	117	52	40	34	38	32
Uruto	10/14/66 ⎬		93	33	33	22	24	30
Karnavito	10/21/66	160	127	47	45	47	42	35
Boira	11/7/66	116	96	31	39	29	29	25
Korpei	11/16/66	231	206	66	78	45	64	50
Sieronji	12/3/66	78	62	13	23	—	—	9
Pomaua	12/8/66	176	115	50	41	28	22	34
Bairima	1/5/67	73	60	22	24	14	16	21
Turungum	2/7/67	118	89	21	23	—	—	19
Moronei	2/14/67	123	114	33	37	—	—	32
Old Siwai	2/20/67	—	26	24	—	—	—	25
Arawa	12/15/66	134	109	26	36	17	20	20
Rorovana II	1/10/67	289	246	85	93	41	58	54
TOTALS		2507	2008	675	697	418	469	526

tometrics, male and female dermatoglyphics, and blood polymorphisms) were eventually analyzed to provide spatial configurations of the village relationships, and the resulting nine configurations were compared to discover which configurations most closely resembled the rest, and which were most unlike.

If all the patterns of biological variation appear to be homogeneous in their broad outlines, then the logical conclusion is that polygenetically determined traits vary much as Fisher thought in natural populations, closely following patterns of genetic heterogeneity. However, congruence of the biological patterns with migrational, linguistic, and geographical relationships would suggest further that the patterns are being determined primarily by sampling events and historical accidents.

Natural selection may be effective or even intense among these village populations, but from the approach I am using here it should be primarily a disruptive element, distorting what constancy there is in organic patterns. The distortion may accentuate differences among groups (differential selection), it may act to keep the populations uniform (stabilizing selection), or it may even act in approximately the same pattern as migration and drift. By treating broad categories of variation at once (seven different blood polymorphisms, measurements from different parts of the body, and so on), I may very well be smoothing out some evidences of selection over this area. But under the circumstances, this is not objectionable, but rather a justifiable course of action.[3] In doing so,

3. Incidentally, I recognize this possible explanation is also a plausible one for the aforementioned constant rates of protein "decay" among related species. There is little question that some sections of proteins are much less open to change than others, and that these invariant areas are often identified with some clearly defined function. Stabilizing selection, acting at the same high intensity to protect the same function over thousands of generations in divergent species, can maintain the same structure. But where structures may be modified without immediate and severe detriment, then, over a long period of time, the corresponding structures in two species may well become increasingly different, rather like some elements of linguistic change. The average difference in protein structure between two species should then become some function of the intensity and uniformity of stabilizing selection and the time of divergence of the species. Selection which is variable in its intensity and direction will only add to the rate of decay.

Of course, different proteins do not diverge in structure at the same rates, almost certainly because some have much more stable functions than others. Histones IV, involved in genetic repression, show an incredible uniformity among higher organisms, while hemoglobins are many times more variable. Undoubtedly the mechanisms involved in genetic

I also hope to elucidate the relationships of these Pacific Islanders in a way that is interesting to the archaeologist and social anthropologist, as well as the human biologist.

---

repression are much more conservative and universal to metazoans than are the fine points of vertebrate respiration.

In measuring change at so many different genes over such long periods, it is inevitable that any forces which act equally over all genes and over the entire time extent will have magnified or predominant effects. I believe, for example, that King and Jukes (1969) have showed that there may well be a higher mutation pressure toward amino acid residues which have multiple DNA codons than toward those with only a single codon. But that is all. They have said very little about the nature of natural selection in proteins.

2 The Setting

To the majority of outsiders who are familiar with the name, Bougainville Island conjures up memories of mosquito-infested mangrove swamps and some of the most difficult jungle warfare of World War II. A newer connection, now common in Australia at least, is with huge copper-mining operations and political turmoil. This newer association may begin to replace those of wartime, depending on how spectacularly the situation in that area deteriorates. But, at the time of this survey in 1966–67 and 1970, in spite of these and earlier incursions from the industrial world, most of the traditional life patterns of the island's native inhabitants remained, and in spite of what the passing soldiers and miners may think, Bougainville's inhabitants are quite extraordinarily heterogeneous, both linguistically and physically. The chapters that follow will demonstrate the extent of the diversity of some of the populations, but here I wish to establish those elements of the environment, its utilization by the people, and their history which have acted to influence the biological variation in these populations in a significant way. I realize that this means presenting a limited account of conditions on the island, but to do otherwise would greatly increase the length of this account. The following remarks are a distillation of a number of different sources which should be consulted for a more detailed description.[1]

Although Bougainville and Buka Islands are currently politically linked with New Guinea as part of the new nation of Papua New Guinea (a United Nations trust territory at the time of the surveys), their geographical and historical ties are with the neighboring islands of the Solomons Archipelago which extends to the southeast. Other than these contacts, which at times have been extensive across the Bougainville Straits, Bougainville and Buka have most likely been remote from persistent and frequent communication and migration with other island areas of the southwestern Pacific. The next nearest island group is over 100 miles to the northwest (the Bismarck Archipelago), although inhabited coral atolls are considerably closer to Buka. Moreover, there is so much that

1. For environmental considerations, see the series of papers in Scott et al. (1967); for linguistic information, Hurd and Allen (n.d.); for medical and epidemiological information, Scragg (1954, 1967), Vines (1970), Damon (1974), and Page et al. (1974); for ethnographic information (particularly extensive for southern Bougainville non-Austronesian speakers), Oliver (1954, 1955, 1973), Ogan (1972), and, where applicable, Mitchell (1972), and Nash (1972); for archaeology, Terrell (1972), and Terrell and Irwin (1972).

is ethnographically distinctive about the two islands that they have generally been treated as a distinct culture area in the Pacific.

This is not to argue that Bougainville and Buka have been totally isolated from population movements except from the southeast, but that possible influences from other areas are not likely to have been major determinants of the biological composition of the general population once it had become established, both because of their infrequency, their small numbers, and the likelihood that the newcomers were closely related to the already resident populations.[2]

Of the Solomons chain, Bougainville is the largest (130 miles long and approximately 40 miles in maximum width, slightly smaller than Puerto Rico), the most populous (the official census estimate contemporary with the initial survey in 1966 was 72,000 inhabitants), and, even disregarding the revenue from the colossal mining operation, by far the wealthiest.

There is no lack of fresh water on Bougainville, except for the north coast area. Every village is located close to some stream or river, of which there are a great many, though none of the rivers exceeds 200 yards in width and a few feet in depth, so that, except for the largest, rather than being significant barriers to human movement, streams and rivers are focal points of human activity in their courses through the hills.

The climate is wet and humid and remarkably uniform throughout the year, and throughout the island. It is true that some low-level areas, such as Kieta township, can be oppressively hot in the heat of the day, while villages at high altitudes are often blanketed by clouds for days. Yet nights everywhere are cool, so that daily variations in temperature everywhere greatly exceed those of means for different months. The average annual temperature at sea level is

2. This argument follows Terrell (1972). "Migrations" do occur, and have their effects, as the Torau movements over the past centuries attest. A less well-known incident, involving a single migration from Buka, had important results for the small atoll group of the Carterets. The few inhabitants of these islands were originally Polynesian. From their low perspective they could see the high mountain peaks of north Bougainville without being seen or detected by the Bougainvillians in their villages over the horizon. A fisherman from Hanahan village on Buka, according to local accounts, was blown out to sea in his canoe, was picked up by men from the Cartarets, entertained on the island, and then sent home. He returned the hospitality by organizing a raiding party from Hanahan which overran the small and surprised Cartaret force, killed all the males, and took the women as their wives.

80 degrees Fahrenheit, and the variation in monthly means is only three degrees. During Douglas Oliver's stay in Siwai, he recorded extreme temperatures of 96 degrees (during several midafternoons during April and May) and 64 degrees (just before sunrise during October). During the period 1960–1965 the mean annual rainfall was 120 inches at Kieta on the east coast and 130 inches at Tonu Mission in Siwai on the west, although Scott et al. (1967) report that it amounts to only 105 inches on the north coast.

The lack of seasonality is probably a major factor in the traditionally high productivity of the plantations of the island. More importantly for this study, the principal subsistence crops may be, and are, planted at any time during the year, and there is no need to grow a surplus for storage. These include above all the sweet potato, some taro, which was the major staple before 1939, and yam, banana, sugar cane, papaya, breadfruit, coconut, and cassava.

Fishing, hunting, wild-plant collecting, and pig-raising have been generally less important to diet (see Oliver 1954, 1955, Ogan 1972, Mitchell 1972). The northern beach-dwellers do fish a good deal, but elsewhere along the coast fishing has always been hazardous and not very productive. Freshwater fishing carried out by trapping, spearing, and arrow-shooting provided little extra food. People also hunted wild pigs, and in the south pig-raising was important as a cultural activity. However, nowhere was there a regular supply of pork, and unlike the reported situation in Highland New Guinea (Rappaport 1967), the raising of pigs, instead of demonstrably increasing the protein and caloric intake of the people, was done at considerable cost and labor, as the pigs do considerable damage to the gardens and seedling coconuts (Ogan 1972).

Diversity within the Island and Its Population
A northern and a southern range of mountains and volcanoes make communications between the east and west coasts difficult in most areas, although hardly impossible (Figure 2.1). The northern range culminates in the active volcanoes Balbi, 8500 feet high, and Bagana, 5700 feet high, while the southern range is generally lower (Plate 1). Earthquakes associated with the vulcanism are about as violent and frequent as anywhere on earth. Associated with this tectonic instability is a very high rate of erosion and deposition of soil.

Only about 6 percent of the total land area, or 200 square miles, is under

2.1 Topography of Bougainville and Buka Islands

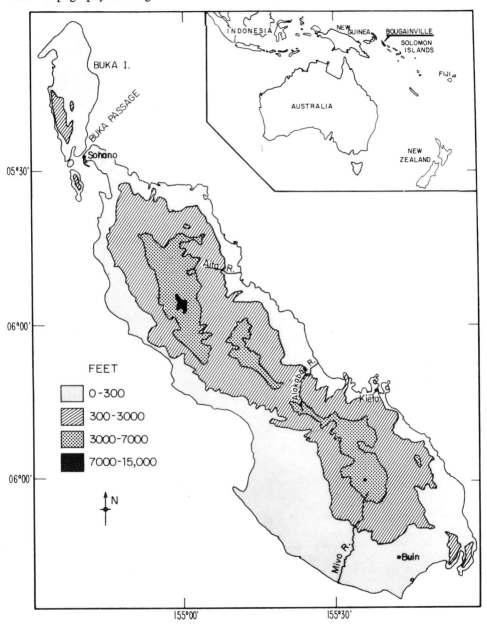

subsistence cultivation, and even though slash-and-burn farming does not require land of high potential yield in terms of mechanized commercial agriculture, there is nevertheless a high population density on nearly all the land most suitable for farming. A few areas with good soil appear to be relatively free of population, principally near Boku on the southwest and Numa Numa plantation in the Rotokas area in the east. The fragmented distribution of the language groups on the island suggests the limiting and divisive effect the terrain has had on the island's settlement. Most languages are spoken over distances no greater than 10 miles, and most of these are subdivided into sublanguages and dialects.

While the population density is sparse in comparison with most other areas

1. The mountains of Bougainville: From Balbi south to Bagana

in the populated world (currently 17 people per square mile for Buka and Bougainville, and about 10 per square mile for Bougainville before the Second World War), the distribution is quite irregular, ranging from 0 to 160 people. The preference everywhere is for land that is well drained, with flat to gently sloping contours (Plate 2). Over half the total population lives on such land, in two quite different locations (Figure 2.2). Most of the people of Bougainville itself are concentrated on the plains of the south, which are separated from direct contact with the coast by a broad belt of swamps and poorly drained ground. As a result, the villages there are strung out midway between the sea and the mountains, in the inland southern hills on both coasts. The other great concentration of people centers along the coasts of Buka and northern Bougain-

2. Cocoanut grove in the hills above Manetai

2.2 Population distribution on Bougainville and Buka Islands

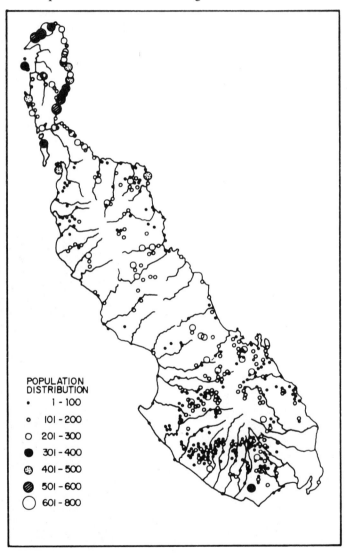

POPULATION
DISTRIBUTION
- ·   1 - 100
- ∘   101 - 200
- ○   201 - 300
- ●   301 - 400
- ⊛   401 - 500
- ◍   501 - 600
- ◯   601 - 800

ville, built up from raised coral beaches, and these people depend more upon the sea for their livelihood.

On the other hand, not only are the highest elevations uninhabited, but there is a relative dearth of people in the central portion of the island, near its narrowest width. On the west, this is partially explained by the sharp transition down Bagana's barren slopes to mangrove swamp, a hostile barrier to habitation. On the east the slopes fall away more gently into swamps, and there is a band of tillable soil between the mountains and the sea, which has not been utilized to what would appear to be its full extent. However, it was noticeable that in this area, the principal focus of the survey, there was a strikingly high incidence of goiter among villagers, which is often associated with iodine-poor volcanic soils heavily leached by rain (see also Plate 19).

Patterns of marriage were also important in encouraging the development of cultural and biological heterogeneity over the island. Everywhere, matrilineal exogamous descent groups prevail. Actual matrilineages, where kinship relations are known, are grouped together into larger exogamous matrilineal sibs, or clans, which all share some origin myth, name, and totem. Oliver states that in the Siwai area in 1938 the simple fact of matrilineal exogamy had the effect of reducing by only about one fifth each person's potential mates. Along with this prohibition, cross-cousin marriage was preferred.[3] Oliver found in Siwai that actual cross-cousin marriages constituted about 5 percent of the total, while marriages between more distant cross-cousins, real or classificatory, amounted to another 25 percent. Such a preference is compatible with two-section, direct exchange systems of marriage, where moieties (or "halves") provide each other with marriage partners from one generation to the next. Ogan (1972) has suggested moieties existed in all the inland groups of south Bougainville, and says that as a more general feature of Nasioi life today, traditional exchange is between two opposing social units. He quotes a Nasioi informant as saying, "We always used to marry that way (direct exchange through cross-cousins), until the priests told us to stop because it was like marrying our sisters. We

3. Cross-cousins are those related to each other through their parents' siblings of the opposite sex; that is, either a father's sister or a mother's brother. This means in a society with matrilineal descent groups that one's cross-cousins will belong to different lineages from oneself.

married that way so land and valuables would stay close. Now that the priests made us stop, we have more land disputes."

Most often, at least in the south, the small hamlets of three to six houses which were the ordinary residential clusters, were based upon a single matrilineal group, with the husbands of its female members and the wives and offspring of its resident male members. In other words, although each matrilineage was usually localized in a hamlet, it was not, strictly speaking, a territorial group, as some of its members (more usually male) would have taken up residence elsewhere—in their wives' matrilineage hamlets (Oliver 1973). As the matrilineages owned corporately all economically valuable land, this had the effect of tying hamlets to a somewhat circumscribed area. The male off-

3. Beach village: Rorovana (1967)

spring of male members who had married outside their village were very likely to find spouses back in their grandmother's hamlet, so that again the emphasis was on localization. Local endogamy, with these restrictions, was favored, so that almost everyone found mates within a radius of five miles of their birthplaces. This was still true at the time of this study (see Chapter 3).

As for polygyny and its potential distorting effects in the gene pool of the following generations (see, for example, Neel 1970), it was certainly permitted in earlier times, but to what extent is not clear. It was called "common" for the Nasioi by Frizzi (1913), acknowledged for the northerners by Blackwood (1931), and termed "hardly commonplace" for the Siwai in 1938 (Oliver 1954). Of 900 families Oliver recorded in the area, 809 were monogamous,

4. Bush line village: Turungum (1967)

71 involved two wives, and the remaining 20 involved from three to eight. Tokura, the one man with eight, was so extraordinary that his singular achievement was still the subject of discussion in the area in 1970. On the other hand, as a result of divorce (or spouse's death) and remarriage, nearly every adult was married to two or more persons during the course of his or her reproductive life.

Political organization, or its lack, further contributed to diversity. If a tribe is taken to be a group where all the members are ordinarily at peace with one another and enjoy access to one another's land, and usually fight as a unified group under a single leader when occasion demands, then groups of hamlets within the same neighborhood most closely fit the term for most of pre-colonial

5. A part of Arawa town (1973)

Bougainville (Oliver 1973). The more cohesive neighborhoods or tribes were to be found among the coastal peoples, and also in the Buin area of the southeast. Elsewhere they appear to have been smaller and more loosely organized. Warfare was ubiquitous, and although cannibalism was certainly practiced in Buka and northern Bougainville, and head-hunting for trophies in the south, the casualties were probably fairly few. Warfare undoubtedly caused many people to shift their houses to less exposed locations, and inevitably limited the exchange of mates from one area to the next.

Taking into account what is known of the ecologies of the island, the current population distribution, the traditional technologies available, and what he has been able to learn from archaeological surveys in north and south Bougainville,

6. The population explosion: Adults and preadolescent children of Kopani village, Eivo

John Terrell (1972) believes he can infer the long-term survival of a number of important population patterns. No one currently wants to be quoted on a date for the earliest settlement of the island, but 5,000 years ago is no longer an unreasonable estimate. Terrell sees evidence for a trading network linking the coastal villages on the northeast shore and Buka existing for at least the last 2,000 years, and these were clearly not the earliest Bougainville settlements. On the other hand, Terrell thinks the limited evidence from the southern plains is enough to suggest a markedly divergent prehistory from the northern shoreline, with each of the four or five population (language) centers there developing very much in its own distinctive manner, all for the most part in isolation from each other and from the rest of the Solomons. By inference, the isolated populations of the northern mountain slopes have an equally long history of development in comparative isolation. It is clear, however, that at least in historical times, significant influences on the more southerly populations has come from the islands across the Bougainville Straits. Also, it is likely that small numbers of these beach people have occupied the narrow strips of habitable coast on the southeast shore for many centuries, still maintaining contacts by sea with the Shortlands.

Language Relationships

The entire situation outlined above undoubtedly contributed directly to the incredible diversity of languages spoken only on Bougainville and Buka. In 1938, and presumably during the previous fifty years at least, there was a total of nineteen languages spoken by the 45,000 or so inhabitants of Bougainville and adjoining Buka. Eleven of these were of the sort labeled "Austronesian," and the other eight "Papuan." Most of these undoubtedly first appeared as new variants of older languages on Bougainville itself. Their familial relationships to other languages on the island, and their lack of relations to any off-island languages, bear this out.

Outside Australia, Austronesian languages cover the South Pacific region, including the aboriginal languages of Taiwan, Polynesian, Malay, Javanese, and even a language in South Vietnam. They are all believed to be historically interrelated (as is English to German, Russian, and even Bengali), as they all share some features of vocabulary and grammar that presumably derive from some parent language. Only the so-called Papuan languages, spoken over most of the New Guinea mainland and on a few of the other islands of western Melanesia, are clearly unrelated to this enormous group. These languages are so diverse themselves that it has been impossible to justify their classification into a single group, although Stephan Wurm has proposed a number of large groupings which cover most of the known languages.

As for Bougainville, the Austronesian languages were, and are, spoken throughout Buka and on or near the Bougainville coasts; the non-Austronesian or Papuan ones on Bougainville alone, and mainly in its inland areas (see Figure 2.3). Using information from word list surveys alone,[4] Allen and Hurd have divided the Austronesian languages into four "families" of the Bougainville Melanesian Stock (a subdivision of Austronesian):

Tinputz family: Tinputz, Teop, Hahon;
Petats family: Petats, Halia, Solos, Saposa;
Banoni family: Banoni and Nagarige-Amun;
Torau family: Torau, Uruava and Papapana.

Most of these "families" are further differentiated into dialects and sub-languages. The indications are that these languages are more closely related to Austronesian (Melanesian) languages to the south in the Solomons than to those of New Britain and New Ireland. Simple and perhaps naive attempts to estimate how far back in time the Melanesian language families of Bougainville began to diverge by the method known as glottochronology (developed from estimates on Indo-European languages, and full of pitfalls), have yielded figures of approximately 4,500 B.P. or even earlier (Terrell, personal communication).

In addition, there is the possibility that these languages represent "creolized" mixtures with Papuan languages (Capell 1962). However, it is not clear that such distinctions are based on linguistic relationships alone, but rather take

4. Allen and Hurd's survey of Bougainville in 1963 was based solely on shared cognate percentages derived from a list of 170 expressions, an expanded version of Wurm's (1960) word list. Using Swadesh's (1955) criteria, "Groups sharing between 93 and 100 percent of their basic vocabularies are said to belong to the same *language;* languages sharing between 28 and 81 percent belong to the same language *family;* those sharing between 12 and 28 percent belong to the same language *stock;* and those sharing between 4 and 12 percent belong to the same language *phylum.* Also, all languages in a given family must have at least 28 percent shared cognates with every other language in the family."

2.3 Language areas, affiliations, and locations of the 18 survey populations

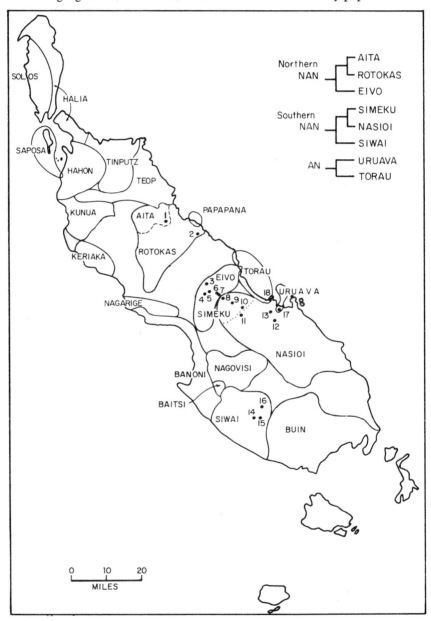

into consideration the distinctive black skin and physical appearance of Bougainvillians and western Solomon Islanders. In a personal communication, Capell wrote he hoped my own findings might help establish relationships of Bougainville Melanesian-speakers with other off-island groups which would benefit the linguists rather than the other way around.

The Papuan languages of the interior are even more distantly related than the Melanesian languages. This fact, along with their interior locations, implies a much older and earlier settlement. They are classed by Allen and Hurd as follows:

NORTHERN BOUGAINVILLE PAPUAN STOCK
Rotokas family: Rotokas and Eivo
Kunua and Keriaka have no family relationships

SOUTHERN BOUGAINVILLE PAPUAN STOCK
Nasioi family: Nasioi and Nagovisi
Terei family: Buin and Siwai

Capell (1962) has drawn distinctions between the grammars of the two different stocks. The northern Papuan cluster has complicated verbal systems but lacks the elaborate noun classification systems characteristic of the southern half of the island. The southern Papuan stock is characterized by the classification of numerals into upwards of 40 classes according to the nature of the objects being counted. These languages also have object-incorporation of the pronoun into the verb. These two stocks of Papuan languages may be placed in a single language phylum, sharing from 4 to 12 percent of common vocabularies. Wurm has proposed recently that they are remotely related to those found in other parts of the Solomon and Bismarck Archipelagoes. No relationships have yet been established with New Guinea, although this too may come.

These various linguistic relationships generally correspond closely to the accepted cultural and historical ties between groups. In the south, the four Papuan-speaking groups hold many culture traits in common, but Oliver (1955) and Ogan (1972) differentiate between the "mountaineers" (Nagovisi and Nasioi) and "plainsmen" (Siwai and Buin). The Siwai and Buin were more open to influence from the more vigorous and technologically advanced Melanesian-speakers from the southern offshore islands, from whom they adopted shell money, pottery, and other cultural features.

Before European control had put an end to warfare (as well as to the extensive trade in this area), the Melanesian-speakers not only traded with the Siwai and Buin, but habitually raided the southern coast for concubines, slaves, and human trophies. Melanesian-speakers occupied sections of the southeastern coast at the time of German pacification early in the twentieth century, but a great many of these people have been assimilated by the Nasioi. All the Uruava now speak Nasioi, but north of Arawa three remaining Torau villages retain their distinctive language.

While the northern Papuan-speakers are the least well-known of the three language stocks on the island, the indications are that like the southerners, they form a distinct cultural unit. These people had, and some still have, very important male initiation rites which were completely lacking in the south.

Physique

Although the people of Bougainville, Buka, and the neighboring western Solomon Islands are easily identifiable from other Pacific populations by their very dark complexion, frizzy hair, and generally African appearance, Western observers, almost from the first, have noted distinctions in the physical appearance of different groups of Bougainvillians. Again, these have reflected, one way or another, the three linguistic and cultural divisions of the population. Guppy, the first Westerner to make anthropological notes of any consequence (1887), noted a "bush/beach" dichotomy existed within the island, and that the shore people he measured in the Bougainville Straits seemed taller and more long-headed than people who inhabited the inland areas. After the Germans established their protectorate over the island in 1884, Parkinson (1907) and others (Ribbe 1903, Schlaginhaufen 1908, Thurnwald 1912, Frizzi 1914) in their reports elaborated upon this theme. After World War I, when Bougainville became part of the Mandated Territory of New Guinea under Australian administration, the full extent of the differentiation within the island became plain. Most of the interested visitors accepted a tripartite subdivision of the Bougainvillians.[5] These were the Melanesian-speakers of the coasts and the northern and southern "bush" people of the interior. The criteria for this separation was primarily cultural and linguistic, with physical considerations playing a secondary role.

While in the midst of his ethnographic study in 1938 and 1939, Oliver measured a large series (over 1300) of mostly young Bougainville males from most of the sections of the island. While approximately half of the subjects were from Siwai and Nagovisi, and most of the others were a select group of plantation workers, this massive amount of information led to a systematic description of the anthropometric and anthroposcopic variation of the islanders (see Oliver 1954, Oliver and Howells 1960, Howells 1966a). The general conclusion was that, again, there were three distinct "types" of physique: small southern inland Papuan-speakers; larger northern mountain Papuan-speakers; and also big, but distinctive, Melanesian coastal peoples (see Plates 13–17).

This physical division into three is also recognized by at least some Bougainville men I interviewed. Two respected leaders, especially, gave similar descriptions of the physical "types" Oliver and Howells found, and which I also saw emerging from the complicated multivariate analysis to be found in Chapter 5. They talked about tall, rather long-headed beach people, big and broad ("thick") and heavy northern mountaineers, and small and short-headed southerners.[6]

The Effects of European Contact on the Populace

To ignore the effects of World War II on the lives of the people of Bougainville is to deny reality. The war years divided the recent history of the island neatly in two, the prewar years being characterized by comparatively modest disruptions of indigenous life patterns. However, there is also little question that

5. Chinnery (1924) made the best record by government officers; Thurnwald (1912), Blackwood (1935), and Oliver (1955) provided excellent ethnographies of particular groups; and Fathers Muller and Rausch made the most voluminous linguistic contributions (see Allen and Hurd n.d.).

6. In passing, I might add a similar story Ernst Mayr tells of New Guineans who correctly identified all the true avian species in their area, with only a single "mistake."

an earlier watershed was the introduction of iron tools with the advent of traders and the infamous "blackbirders" in the latter half of the nineteenth century. This has led Ogan (1972) to differentiate between what he calls "aboriginal" and "traditional" society on the island, "traditional" referring to the comparatively stable conditions existing from the time of first contact with industrialized societies to the Second World War. Metal tools opened up new areas of fertile land covered with only secondary vegetation, which made possible the accumulation of wealth in foodstuffs through individual effort.

To Ogan's mind, the wealthy and powerful "Big Man" of Melanesian ethnographic fame was a phenomenon of particular importance in "traditional" times in Bougainville. Before then, at least in the Nasioi area, social roles may have

7. Christmas "party" at Bairima village, Nasioi (1966)

been more homogeneous, with relatively equal distribution of land and wealth among families. This is an important point concerning fertility patterns, suggesting there were no powerful men with favored procreative status in aboriginal times.

The impact of Western culture was varied in its nature and intensity on different parts of the island during the prewar period (see Oliver 1949). Traders and labor recruiters, such as R. Parkinson from New Britain, had been visiting the island during the late nineteenth century, but the first major and permanent contact with Europeans (predominantly French and Germans) came about in the Kieta area with the establishment of a Roman Catholic Mission in 1902 under the sponsorship of the Society of Mary, moving in from the British Solo-

8. Shopping at the Arawa Supermarket (1973)

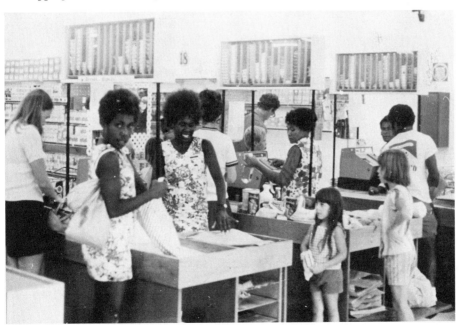

mon Islands (Frizzi 1914, Parkinson 1907). Although the natives drove the missionaries out at first (White 1965), they returned before the government of Germany established an outpost at Kieta in 1906.

According to all accounts, the early missionaries left the bulk of native life untouched, except for a few customs which were particularly abhorrent to them. This seems to have been a deliberate policy of Catholic missionaries generally; they seem to have been aware that, by demanding drastic changes, they might bring massive social disorganization to their converts along with the gospel. They instituted baptism and other sacraments, debunked the older supernatural beliefs of the natives, and attempted to change marriage customs and almost all institutions relating to sexual functioning. Also, and importantly for this study,

9. Meeting place: Tunuru mission (1967)

missionaries were strongly opposed to polygyny and cousin marriage, and also divorce, child betrothal, and the celebration of a girl's menarche. Priests claim that they also halted the practice of ignoring the least desirable of a twin pair and letting it starve, but some ethnographers, Ogan, for example (1972), are not sure how widely this practice prevailed in the first place. In any case, the overwhelming majority were converted to Catholicism, with Methodists and a few Seventh Day Adventist groups scattered around the island.

Mission influence has tended more and more to radiate from administrative centers on the island, and the missionaries have often acted as initiators of administration policy in the more remote areas, extracting assistance and co-operation in turn from the government. This is not to say that the Catholic mission has been the tool of the administration, for increasingly, as the administration has become more powerful, there have been sharp differences of policy between the two. Before 1960, the presence of the Catholic mission was nearly overwhelming. Approximately 90 percent of the schools and a smaller percentage of the health facilities on Bougainville were operated and maintained by the missions (see Plates 10–12).

Ogan (1972) says it is likely that up until World War II education as it was dispensed by the missions consisted for the most part in preparing a child for first communion and confirmation, teaching boys to use European tools in semi-skilled tasks, and inculcating some elementary hygiene in both sexes. Except for the initial period of German administration prior to World War I, Pidgin English was taught universally at these schools, as the use of English was not encouraged before the Second World War. It may be inferred that the missions tended to break down the isolation of the scattered hamlets of the interior of the island because attendance at school and Sunday Mass increased contact among parishioners from widely dispersed residences and, in many instances, different language areas (see Plates 7 and 9).

The influence of the German and Australian administrations was not so pervasive. By far the most important consequence for present concerns was the pacification of all the island by 1936, except for parts of Eivo and Aita as well as parts of the west coast. Many hamlets were also consolidated into "line" villages under the Australians.

The plantations have been a third major agent of westernization on the

10. Agent of modernization:
Sister Marie Augusta, Tunuru Mission

island. These are mainly located along the northeast coast, but young men from most areas of the island come there to work for a year or two, and it is hard to estimate the extent of the influence which this stay has had on Bougainville society over the last seventy years. Almost all of the men return to their villages. A few workers from New Guinea and other islands have stayed in Bougainville and taken local wives (see Table 3.8). Certainly the plantation experience has done little to convince the local populace that whites (at least Australian overseers and plantation owners) had much consideration for their welfare. Some sources state that the long-term familiarity of the Bougainvillians with plantations has caused them to be suspicious or hostile toward whites,

11. Baptism at Manetai mission: Father Nicholas Kutulas

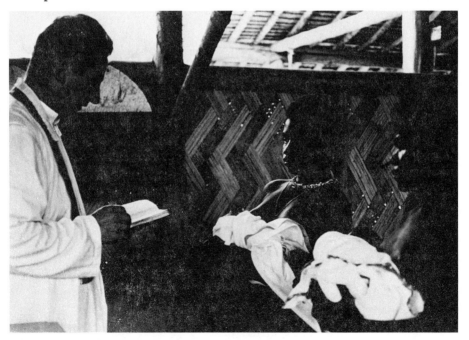

while the lack of exposure to a plantation experience accounts for the relatively eager and docile attitude of the Highland New Guinean.

The Second World War was a watershed in changing social and cultural patterns on the island. The Japanese established garrisons at Kieta, Buka, and at Buin in the south, which, together with some smaller outposts around the island, probably numbered 65,000 men in all (Long 1963). As these men were soon cut off from their sources of supply by the highly effective Allied submarine activities in the Pacific, the initially salubrious occupation policy of the Japanese turned into one characterized by the confiscation of food and the institution of forced labor. Natives and Australians claim that in the last days of the occupation some of the starving noncommissioned soldiers resorted to can-

12. Newborn Bougainvillians and their nurses

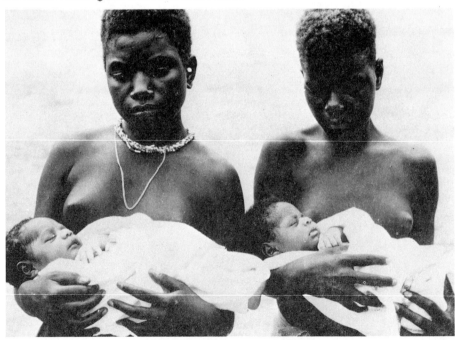

nibalism, preferring the hindquarters of young native boys. Prior to the American landing on the west coast at Empress Augusta Bay, bombing raids were carried out, not only aimed at the Japanese base at Buin, but evidently rather indiscriminately over native hamlets and villages. The bombs did not kill many people but they did demoralize them. Ogan's Nasioi informants, and some of the Siwai subjects of this survey on the west, say they spent much of the year 1943–44 living in the bush away from their homes, "like wild pigs." More remote areas, such as Eivo, suffered little.

It is generally agreed that people suffered more casualties from malnutrition and exposure than from direct military action, although fighting between Australians and Japanese continued, principally in the Siwai and Buin areas north of Buin town, until the final unconditional surrender. There is reason to believe that infanticide, or the abandonment of infants, was a conscious and frequent choice of many couples during this period of hardship. All these effects had a serious net result which is plain in the 1967 population pyramid (Figure 3.2). Note the severe "waisting" in the 20–29 year categories, which is more marked than the corresponding (and slightly earlier) phenomenon in the 1965 U.S. pyramid.

As for permanent disruption of village and language group residence patterns in the affected areas, people did not return immediately to their garden plots and line villages.[7] Particularly in the combat zones, the war experience and its aftermath provided a profound psychological and social trauma. Many of the survey subjects said they had gone to the Allied enclave at Empress Augusta Bay, sick and starving, to be met with seemingly miraculous piles of food towering over their heads. For some areas, the end of hostilities meant the restructuring of the line villages on a grander scale. Mitchell (1972) reports that relatively huge paramilitary villages (population 700) appeared in Nagovisi. In more remote areas such as Eivo, there is no evidence of such phenomena at that time. Even where the fighting and disruption had been greatest, the ties to owned garden plots seem to have pulled people back to their prewar homes. In northeast Siwai, where Oliver had carried out genealogical and census surveys in 1938–39, I was able to establish that, almost without exception, those

7. This statement is based on a comparison of Oliver's records and my own follow-up inquiries in Siwai.

survivors of the holocaust eventually had returned to their own land plots and their old village vicinities. Very few had settled more than ten miles from their prewar houses. Some of the old men said that even during the heat of the battle they used to return from the bush at night to survey their villages and plots, never really letting go. Such is the power of property.

Shifting patterns of residence over short distances continue today. Although in Nagovisi the large villages of the immediate postwar era broke up in the 1950s, the Eivo now are trying to decide if they should all join in a single village of about 900 under the leadership of one strong leader. Up to this point, his efforts have been resisted, largely because people are not ready to move too far from their gardens.

In sum, then, residence patterns have continued to shift over time, the fission and fusion processes[8] really never ceasing to this day. Yet, in all this Heraclitan flux, the amount of large-scale movement seems to be remarkably small. Unquestionably, homogenizing is currently under way, with increasing intervillage contacts and local government councils. But in the survey region in 1967, the expected increase in out-migrations and out-marriages was still small and in the Siwai area, the percentage of marriages between Siwai and Buin did not seem to have increased significantly since before World War II.

Since the war, government influence and control have been more pronounced. Still, according to Ogan (1972), "out of sight, out of mind" was the byword of the governing of Bougainville as much as any more grandiose theory of colonial rule. A common tale passed around the Kieta Pub had it that Territorial maps in Port Moresby (the administrative capital) never included Bougainville before 1950. The situation changed rapidly after 1964 with the discovery of large copper deposits.

Administrative efforts have often suffered from a lack of personnel. The "development" of native agriculture, like other forms of directed social change, received much greater attention after the war, but still the first full-time agricultural officer was not assigned to Kieta until 1958. Since then these "didiman" have tried to introduce cash cropping on a large scale. These were largely unsuccessful and sporadic attempts at first, but have recently become viable and

8. This is a term first popularized by Fortes (1949) and applied to microevolutionary studies by Neel and Salzano (1966).

very important in some areas able to market their produce. It is not at all clear, however, that this pattern can continue to expand for long because of limited arable land (see Mitchell 1972).

In the field of public health the government had little impact before the war. Public health records from that period are meager. Mair (1948) noted that a native hospital was established in Kieta in 1922, and some coastal Nasioi were trained as medical orderlies in the thirties. Nasioi informants of Ogan's mention hygienic measures directed by patrol officers—for example, the digging of latrines and the maintenance of houses in reasonable states of repair. In the more isolated areas (to the north in my sample survey), such patrols were even less evident. As with some other areas of concern—education, for example—responsibility all too often devolved on the missions by default.

The postwar years have been marked by greatly increased attention to village health. The medical staffs were increased by European emigrés who were allowed to practice in the Territory relatively freely, whereas they faced a number of obstacles to full accreditation in Australia. Until 1964 the Kieta Hospital was in the hands of a white medical assistant, at which time a medical doctor (with assistants) became the staff head. An effective WHO anti-yaws program was instituted, but by far the most important innovation was the Malaria Eradication Programme. In 1960, spraying village buildings with DDT began, and mass drug administration was initiated in 1962, again centering around Kieta. Parasitological tests carried out by the government note a sharp reduction in the parasite rate between 1962 and 1964 in the Nasioi area. It is undeniable that the program has effectively eliminated malaria as a major killer (see Chapter 3).

The sharp reduction in infant mortality, the great increase in the number of surviving children in the ten years preceding 1966, and the booming population explosion (see Plate 6) are all attributable in very large part to the effective control of malaria. Also, an infant and maternal welfare program commenced in the area in 1965, and more and more mothers are beginning to have their babies at mission or government aid stations and hospitals. I suspect that the marked secular increase in size of late adolescent males (see Chapter 5) may reflect the improved health conditions as much as any improved nutritional standards.

The overriding new source of Western influence only beginning to make itself felt during the period of the surveys was the effort to mine the large low-grade copper and gold deposits in the central mountain area in the north Nasioi (Guava) region. Under the New Guinea Mining Ordinance of 1922, all mineral rights were vested, not in the landholders, but in the state. After an initial period of apparent cooperation on the part of the local villagers, the exploratory mining phase met with increasing resistance. After conferences and explanations, the miners continued with only official approval. By 1966 opposition by landholders had grown to the extent that a real potential for violence existed, and a police detachment remained on permanent duty in the area. An amendment to the mining ordinance then in effect, introduced by a Member of the House of Assembly from Bougainville and supported by the Catholic mission, was passed in 1966 so that the Nasioi landholders are guaranteed a small royalty as compensation for the destruction of their subsistence. This measure has not pacified those who claim landownership in the area. Many of them say they do not want any settlement except one in which they retain their land and in which no mining is permitted, a most unlikely outcome.

Beyond the immediate area, the effect of the proposed mining operation was already great by 1970. New roads attracted Bougainville laborers and truck drivers, as well as a large contingent of Highland New Guineans. Villages close to the roads (including Korpei, group 10 in the sample) have tended to move toward them. One village, Rorovana (no. 18), has been forced to sell a large tract of land to the mining company for the development of a port facility. Arawa (no. 17) has, by this time, become an appendage of the planned company town of the same name for 10,000 non-indigines. This town of Arawa will therefore become the second largest in Papua New Guinea, surpassing Lae and Rabaul in size (see Plates 5 and 8).

The parent mining company, Conzinc Riotinto of Australia, was increasingly compelled to chart policies independent of those recommended by the Australian administration, which at the beginning led them into great difficulties with the Nasioi landowners and threatened to bring all mining to an end. Some elements in the upper levels of the company hierarchy, recognizing the self-government was a strong and imminent possibility, made serious efforts at accommodation with at least the major political forces on the island. Salaries have consistently been raised from the initial low levels, which were comparable to plantation wages. The mine is sponsoring a new technical school, a radio station, and is responsible for new wharves, roads, and medical facilities. And the company has gone so far as to retain an anthropologist as a consultant.

The Australian administration, it is generally agreed, made serious mistakes in its early opinions, but, more fundamentally, its aims are different from the rather simple profit motive of the miners. As a senior administrative official said in 1966, "Our principal obligation is to alleviate the tax burden of the Australian taxpayer," a statement which implied that mining royalties should go toward financing the Territorial Government Budget, and not to the general Bougainville population or any small segment of land claimants.

The mission initially took a different tack, increasingly opposing the mining operation as it developed and supporting what it saw as the rights of the natives. A good number of the priests on the island are already indigenous to Bougainville, and the majority of the remainder are conscious both of the attempt to relinquish control to local priests (as one priest said, "We know what happened to the priests and nuns in the Congo who were identified with the Belgian miners") and of the threat to the authority of the church from the mining operation in morals and leadership. All priests complain that the miners have influenced their parishioners to drink more, and to curse, steal, and fight. The brightest students in the mission schools are now more likely to take well-paying jobs in Bougainville Copper than to become teachers or priests, as was the case before.

The mine is already leading to truly revolutionary changes in those areas close to the mining operation. These will most likely take the form of the urbanization of villages, their incorporation into the new towns or roadside slums, increasingly disjunct relations within the villages, but with a peasant hinterland still relatively unchanged. This is not a pleasant prospect, but hardly an uncommon one in underdeveloped countries around the globe today.[9]

The Villages of the Survey

With this complex situation in mind at the outset, I wanted to maximize the language variation in the populations I surveyed occupying the smallest possible geographic area, as mentioned in the Introduction. The east central

9. In his most recent book, Oliver (1973) gives an informed and up-to-date account of the political, social, and economic situations in Bougainville.

13. Man from Nupatoro, Aita, with periodontal inflammation

14. Atamo woman, with decorations

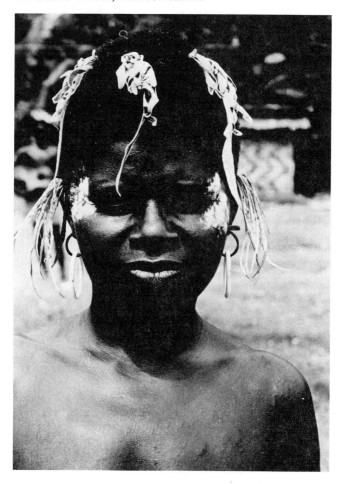

15. Ikirai of Moronei, Siwai, making "Buka Basket"

coast of the island was the obvious choice, and I was able to include people who spoke languages belonging to all three language stocks (see Figure 2.3). The language relationships of the different villages appear in Table 2.1, and their approximate geographic positions in Figure 2.3.

Quite obviously, things are not so simple as they appear on the map. While for most of its history the island's populace has been broken up into small and relatively isolated pockets, with only minimal communications and networks of exchange extending over very short distances, this is not to deny that people, and even entire hamlets and villages, changed locations, but that such movements have been almost always over very small distances. And especially near

16. Peter Tumare of Moronei, making traditional pots in front of his club house (see slit gong)

17. Rorovana man calling people to prayers

Arawa, truly extraordinary changes are now overtaking and obliterating the situation this survey describes.

The "villages" are generally postwar aggregations of older matrilineally based neighboring hamlets (see Plates 3 & 4), although group 16, "Old Siwai," consists of 25 older men from a group of northeast Siwai villages who were included because they had been measured by Oliver in 1938–39 and provided for a longitudinal study of aging and the secular trend (see Chapter 5). As for group 17, "Arawa," only a few elderly individuals there speak Uruava, a Melanesian language related to Torau. The common tongue is the Nasioi of their neighbors, with whom they have intermixed substantially. The Torau village of Rorovana has had recent ties with the islands of the Bougainville Straits (Wheeler 1908, Terrell and Irwin 1972), but to what extent they represent a single discrete "migration" from the south is open to debate.

18. Albino Eivo girl

The Southern Papuan stock is represented by six nominally Nasioi villages besides the two Siwai villages and the one amalgam of 25 men. Actually, three of the villages (groups 8–10, the Simeku) speak a "sublanguage" of Nasioi, having only 67 percent shared cognates with the central dialect and sharing a smaller percentage with the Eivo, their northern Papuan neighbors. Another village, number 11, shares 90 percent of its cognates with the Nasioi core dialect, but this distinction is of minimal importance. As for the northern Papuan stock, the Rotokas (group 2) and Eivo (3–7) belong to the same family. Aita, a "dialect" of Rotokas (with 81 percent shared cognates), is also represented (group 1), and it is possible that this was a much more distinctive language before pacification in the 1950s (Pelletier, personal communication).

Table 2.2 gives the shared cognate percentages of the language groups

19. Goitrous Eivo woman

Table 2.1
Language Relationships of the Sampled Villages

Northern Papuan Stock
    Rotokas language
        1. Nupatoro (Aita dialect)
        2. Okowopaia
    Eivo language
        3. Kopani
        4. Kopikiri
        5. Nasiwoiwa
        6. Atamo
        7. Uruto
Southern Papuan Stock
    Simeku (sublanguage of Nasioi, mixed with Eivo)
        8. Karnavito
        9. Boira
      10. Korpei
    Nasioi language
      11. Sieronji
      12. Pomaua
      13. Bairima
    Siwai language
      14. Turungum
      15. Moronei
      16. "Old Siwai"
Melanesian Stock
    Uruava language
      17. Arawa
    Torau language
      18. Rorovana II

represented in the survey. By performing a principal coordinates analysis of these values, or rather their complements, their relationships can be graphically illustrated, with little lost information, in three dimensions (see Figure 2.4). There is nothing surprising about this picture, and it is, of course, reminiscent of the two-dimensional array of the villages on the ground. The following chapters will compare the language and geographic arrays of the villages with their migrational relationships and patterns of biological variation.

10. See Appendix for a definition of this statistic.

Table 2.2
Shared Cognate Percentages of the Linguistic Groups in the Sample

	Aita	Rotokas	Eivo	Simeku	Nasioi	Siwai	Uruava
Rotokas	.81						
Eivo	.35[a]	.35					
Simeku	.21[a]	.21[a]	.42[a]				
Nasioi	.07[a]	.07	.09	.67			
Siwai	.06[a]	.06	.11	.20[a]	.27		
Uruava	.04[a]	.04[a]	.11[a]	.11[a]	.11[a]	.07[a]	
Torau	.04[a]	.04	.11	.11[a]	.11	.07	.30[a]

SOURCE: From Allen and Hurd, n.d. publ. 1963:21.

[a] Indicates an estimate derived from Allen and Hurd's figures, which usually do not include comparisons of dialects from one language with other language groups.

2.4 Relationships of the survey languages: first and second principal coordinates derived from complements of shared cognate percentages

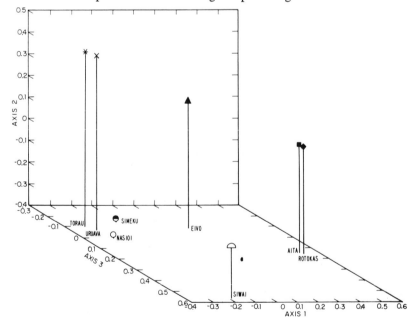

3 Demography of the Sample

The previous section has necessarily been an impressionistic account of the history and shaping forces which have molded the population of Bougainville island and, in particular, the east central region. Here I present as much quantitative information pertaining to the important elements of the demographic picture of the sample villages as I could garner. I fully admit the inadequacy of the data by the standards of Western demographers, but they seem to me to be singularly interesting results, and among the best available for groups without decent hospital or census records (see, for example, Howell 1973).

Epidemiology

I have very few direct data on dietary and disease patterns specific to this sample population and have had to rely upon published and unpublished generalizations or studies from other areas of Bougainville and Melanesia.

Concerning the prewar and pre-contact situation, very little is certain. Malaria was a major cause of infant mortality and adult morbidity. The role of respiratory infections, particularly pneumonia, has been a major one in adult mortality for as long as hospitals have existed in the Territory of New Guinea, having always been the main single cause of death among hospital admissions. Even today it sometimes reaches epidemic proportions. Cancer is now second only to pneumonia as a cause of death in Papua New Guinea's hospital records, the main predisposing factors being chronic skin ulceration, betal nut and lime chewing causing lesions of the cheek, and solar radiation. The most common maladies of minor effect were, and are, such skin diseases as tinea, scabies, and tropical ulcers.

Prior to the Second World War, health services had brought about little change in rural mortality in the entire Territory. Morbidity in some respects had been reduced, but generally speaking mortality rates remained about the same, or were largely uncontrolled. Depopulation of a number of areas is known for the prewar period, with Scragg's (1967) long-term study of New Ireland providing the best documentation. No such massive events of depopulation are known for Bougainville before the war. During the 1950s, rural health workers began using penicillin to combat yaws, sulphonamides, and efficient antimalarials, and initiated immunization programs against tetanus, whooping cough, and tuberculosis. New roads made these services more widely available, and DDT was used to reduce transmission of the prime endemic disease, malaria. The first malaria control program commenced in 1957 in and around Maprik in the mosquito-infested Sepik District. Thereafter pilot projects were set up elsewhere. In 1959 antimalaria activities began in selected areas of Bougainville and New Ireland, and subsequently these activities extended throughout Bougainville to reach almost total coverage by 1961. The effects in reducing mortality, as described below, were dramatic.

Certainly the best epidemiological survey over the years in the Territory has been Scragg's efforts in the Bismark Archipelago, Lemanuka and Solas in north Bougainville (Buka Island), and Tigak and Tabar in New Ireland (Scragg 1967). These sample areas have been followed continuously for about nineteen years, all deaths in the censused population being recorded by age, sex, and cause. For his sample Scragg found that female age-specific death rates were consistently higher than those for males during the childbearing ages only, as implied in the Eivo and Simeku population pyramids presented in Figure 3.3 and Table 3.1. He states that where previously one third of all children born alive in his study died before the age of five, now under 8 percent die by that age, and half of those born can be expected to survive until about the age of fifty. With current mortality, the expectation of life, previously thirty-four years, is approximately fifty-five.

Nutrition

The traditional diet over most of Bougainville, according to Oliver (1954), was 90 percent starch by weight, with taro being the principle source. In 1938 a taro blight forced gardeners to intensify their production of sweet potatoes, which today dominate the diet. Generally the prewar diet was considered to be adequate in terms of caloric intake, and after the starvation conditions that prevailed during the war years, caloric intake returned to and exceeded earlier levels. But according to Emmanuel and Biddulph (1967), protein, calcium, and iodine intake is still probably a good deal less than desirable.

As is the case throughout Papua New Guinea, people residing in villages close to administrative posts and towns often supplement their traditionally starchy diets with sweet and salted biscuits, rice, canned meat and fish, and, among increasingly large numbers of the young men, alcoholic beverages. Vines (1970) reports the consistent association throughout the territory of urbanization (as identified by store-bought foods, emigrant status, and so on) with greater skinfold thickness, higher serum cholesterol levels, higher serum albumin levels, lower serum globulin levels, and lower height-to-weight ratios than villagers. In one of the more systematic nutritional studies, Hamilton and Wilson (1957) found the consumption of protein from purchased foods in the Tolai village of Malaguna reached almost twice the level of that in the diet of the non-urbanized New Guinea village of Busama.

The lower serum globulin levels of urbanized groups in Papua New Guinea have suggested to Vines that their total infection burden is lower than that of villagers in more or less traditional settings. This is a position which runs counter to a large number of anthropological reports of comparatively healthy-looking people in pre-contact situations being transformed into disease-ridden and degenerate victims of Western disease and technology. In a number of such instances it seems to be true, however unromantic, that mortality and morbidity are almost inevitably reduced when such traditional societies are exposed to Western medicine and Western dietary practices. Although African pygmies and bushmen in the bush unquestionably "look" healthier than their cousins on mission stations, medical evidence suggests that, at least in terms of disease, the opposite is true (Sgaramella-Zonta and Harpending, personal communications).

On the other hand, we are introducing many Western constitutional problems along with antibiotics. Increasing urbanization in Bougainville, as elsewhere, will result in increased fatness, increased serum cholesterol levels, and a decreased height-to-weight ratio. Rhoads (1972) showed, in a comparison of factor analyses of body measurements of Solomon Islanders and Americans, that while some Americans are thin and some are fat, Solomon Islanders are never excessively adipose. It could be that blood pressure patterns, with age and the incidence of degenerative vascular disease, will also change.

Bougainville Demography

As people on the island traditionally kept no personal records of births, birthdates, or deaths, it is fortunate that the Catholic mission has shown its customary interest in such matters, and that, at least from the Eivo area to north Nasioi (Manetai and Tunuru parishes), these records for the past forty years are intact. They are not without error. Records in Manetai Parish (the Eivo and Karnavito village in Simeku) before 1930 are obvious estimates for births. Wartime birth dates and death dates were estimated after 1945 when priests returned and uncovered their hidden record books. Another major source of incompleteness and error I believe to be the migrants into the parish, whose ages were estimated by the priest or, more often, by myself. Nevertheless, the results are still more instructive than those from some other widely discussed surveys of tropical gardeners and hunter-gatherers, or from the meager efforts of the Australian administration census in Bougainville, where the usual procedure has been to lump adults into two estimated groups, those aged 16–45 and those over 45.

Population Pyramid

Figure 3.1 and Table 3.1 present the composite age structure of the Eivo, Simeku, Nasioi, and Torau (including Uruava) villages which were covered in the survey in 1966 and 1967. The Siwai, a large proportion of whom are Methodists, and the Rotokas and Aita, do not have adequate records for any

3.1   1967 Population Pyramid of 13 Bougainville survey villages

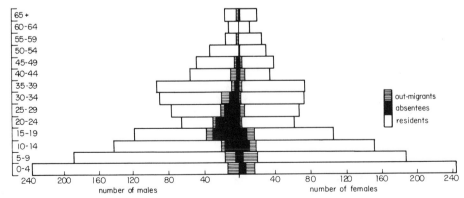

Table 3.1
The Bougainville Population, January 1967

| | Males | | | | Females | | | | Totals | |
Age	Eivo	Simeku	Nasioi	Torau	Eivo	Simeku	Nasioi	Torau	Males	Females
0–4	93	70	34	36	106	69	32	38	233	245
5–9	80	39	29	37	70	52	33	32	185	187
10–14	54	42	23	21	62	27	21	42	140	152
15–19	34	27	26	30	25	36	17	28	117	106
20–24	27	21	9	8	26	20	6	10	65	62
25–29	23	29	12	12	22	26	12	7	76	67
30–34	42	21	14	13	39	14	13	8	90	74
35–39	46	21	16	10	25	20	12	18	93	75
40–44	13	14	14	13	11	7	4	12	54	34
45–49	23	6	8	11	11	9	8	12	48	40
50–54	11	5	7	10	8	11	5	7	33	31
55–59	10	1	2	2	9	4	11	2	15	26
60+	2	5	10	11	14	7	5	9	28	35

demographic estimates. The contrast between the area composite population pyramid for 1967 and the U.S. pyramid for 1965 reveals some important points about the Bougainville situation (Figure 3.2). Bougainville is experiencing a severe population explosion. The Bougainville pyramid has a very large base and small peak—that is, a disproportionate number of children to adults. The median age, in fact, is only 16.0 years, compared to the U.S. median of 27.9. This is a profile which is typically found in groups that are out of equilibrium and are rapidly expanding their numbers, having experienced a rapid decline in death rate (particularly in the early years) coupled with the maintenance of a high birth rate. Current Indian population pyramids look much the same in these respects.

There is a marked "waisting" of the Bougainville pyramid in the 20–24 and 25–29 year cohorts. The people in these groups were infants and children during the Second World War and suffered the most from sickness and hardship. Recall, too, the reports that during this period many couples resorted to infanticide. High rates of absenteeism and migration have been taken into account so that these do not provide a likely alternative explanation. There is a comparable, though much smaller, indentation in the American pyramid, which began somewhat earlier (the 35–39 year cohort was born during the years 1926–1930, spanning the beginning of the Depression) and ended earlier (the 20–24 year cohort, the transitional one on the American pyramid, spans the war years of 1941–1945). In the cohorts over forty years of age there is a tendency for the males to outnumber the females at the base, but to decrease at a more rapid rate through the following five cohorts. This is not a significant difference, but if it is not simply an artifact of sampling error, it may reflect the likely situation existing before the war wherein female mortality would have been high during the reproductive years, but then would decrease relative to male mortality in the postreproductive ages. Certainly there is no evidence for such a difference in mortality in the younger age groups of the Bougainville pyramid, and the small excess of females in the early years, although again taken from a rather small sample, is the opposite of what is ordinarily found in human populations.

3.2   Comparison of Bougainville survey and U.S.A. population pyramids

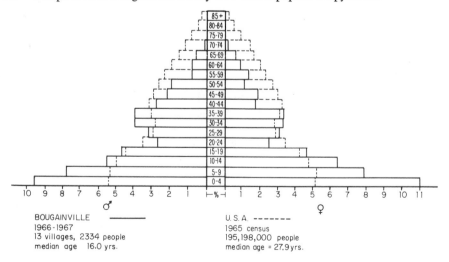

BOUGAINVILLE ———
1966-1967
13 villages, 2334 people
median age   16.0 yrs.

U.S.A. - - - - - - -
1965 census
195,198,000 people
median age = 27.9 yrs.

Turning to the separate pyramids of the villages grouped by language affiliation (Figure 3.3), a fairly distinct dichotomy separates the Eivo and Simeku samples from the Nasioi- and Melanesian-speaking Torau and Uruava. The Eivo and Simeku, numerically the largest groups in the sample (and hence contributing most to the composite picture) have extraordinarily wide bases, fairly narrow waists during the war years, and distinctly different male and female patterns in the postreproductive years. The Nasioi- and Melanesian-

3.3 Population pyramids of villages grouped by language affiliation: Bougainville, 1967

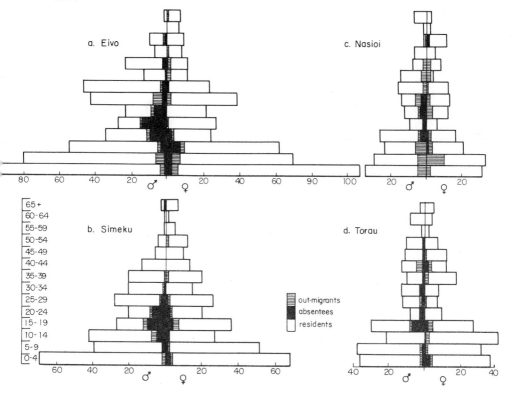

speaking villages, to the contrary, seem to have two distinct elements—a prewar static population, a very sharp wartime contraction, then a very sharp postwar expansion evident in the 15–19 year cohort, with no precipitous increases since that time. These last two groups have been much closer to the centers of westernization (Kieta town and Tunuru Mission) throughout the last forty years than have the more northerly Eivo and Simeku, and the population explosion they experienced shortly after the wartime blight was relatively severe but seems to have tapered off, while in Eivo and Simeku it began more slowly but continues unabated.

There are, of course, other possible explanations for the peculiarities of these pyramids. The small size and errors of the samples make the inferences from the language-group pyramids hazardous. The differences in the pyramids might be due to different in- and out-migration or absentee rates, but as far as these can be ascertained they are not strikingly different. There do seem to be more male absentees in the Eivo and Simeku villages in the 25–29 year categories, but in all four language groups approximately 30 percent of the 20–24 year-olds are off working and return intermittently. The 15–19 year-old absentees are mainly school children. Known out-migrants to other language areas or islands are fairly evenly distributed among the age classes, with proportionately smaller groups in the 0–4 year categories, hardly a surprising finding.

A last point concerns the elderly. There clearly are substantial numbers of people over sixty-five in all four language groups. It is especially likely that these estimates are fairly good for the Nasioi- and Melanesian-speaking villages, those being covered in the old Tunuru Mission records. In a number of particular cases I had photographs, as well as independent age estimates, of some of the old men Douglas Oliver measured and photographed in 1938–39. Some of these men had gray hair and only a few teeth (see photographs), and could not have been less than forty years old at that time. Nobody seems close to being 100 years old, but quite respectable ages in Western terms are not uncommon here.

Crude Birth and Death Rates, and Natural Increases for the Period 1956– 1966

The available baptismal and death records from the mission (corroborated by the government census estimates, which were available for 1965 only) are compatible with the features of the population pyramid.[1] I chose to analyze only the records for the eleven years preceding the survey because these would be by far the most accurate and because I suspected these records would show dramatic changes in birth and death rates.

I expected malaria spraying and other public health measures to affect the death rate, and the records available for the Eivo villages in the survey (Manetai, Table 3.2a) show a dramatic reduction in both the absolute number of deaths and the crude death rate over the period 1960–1964, when the crude death rate dropped from twenty per thousand, to less than ten per thousand.

However, the Torau, Arawas, Nasioi, and neighboring Simeku villages belonging to the Tunuru parish do not show such a marked transition in their death statistics over the past eleven years, consistent with their different population pyramid profiles (Table 3.2b). There is no decrease in absolute number of deaths over this eleven-year period, although the crude death rate does decrease relative to the increasing population.

I attempted to construct age-specific death rates for these eleven years, but because of the small number of deaths in the age categories above five, it proved a hopeless task. Even when averaging death rates for the five years preceding and succeeding 1961, I was unable to detect any differences in the death rates of the post-five-year cohorts. However, Table 3.3 gives the age-specific death rates for the zero to four year-olds for these eleven years in Manetai and Tunuru parishes, and it immediately becomes apparent that 1961 was an unbelievably dramatic year in the reduction of infantile mortality for the Eivo. The number of deaths fell from the previous five-year average of thirteen to three, a figure which was not surpassed in the records, even though the infantile population was growing rapidly. The estimated cohort-specific death rate dropped even more precipitously, from close to one out of every ten children per year in 1956 to less than one out of every hundred in 1966. Another way of viewing the same figures is that, at the 1956 rate, more than one third of the children born would die before they reached the age of five, while at the 1966 rate, less than one out of twenty children born would die before age

---

1. These records are necessarily underestimates of births and infant deaths, as baptisms of the newborn often do not occur until two weeks after birth.

Table 3.2a
Manetai Parish Sample Death Rate.[a] Eivo Plus Karnavito Villages

Year	Deaths	Population	Crude death rate (deaths per 1000)
1956	23	635	36
1957	16	654	24
1958	21	685	31
1959	24	716	34
1960	17	744	23
1961	13	784	17
1962	10	824	12
1963	11	859 (877)[b]	13
1964	4	922	4
1965	8	969	8
1966	5	1024	5

[a] In these and the following tables, the 1966 census figures are taken from my own checking of the mission records. The population figures for earlier years, such as 1965, were estimated crudely by subtracting the difference between births and deaths for 1965 from the 1966 figures, and so forth.

[b] Government census estimate in 1963, given as an independent estimate.

Table 3.2b
Tunuru Parish Sample Death Rate. Simeku [Without no. 8], Nasioi, Arawa and Rorovana Samples

Year	Deaths	Population	Crude death rate (deaths per 1000)
1956	4	757	5
1957	15	779	19
1958	5	810	6
1959	6	842	7
1960	10	868	12
1961	5	912	5
1962	14	942	15
1963	6	988 (1097)	6
1964	13	1030	13
1965	4	1066	4
1966	4	1136	4

Table 3.3
Death Rate (0–4 Year-Olds)

Year	Manetai Parish			Tunuru Parish		
	Deaths	Population (0–4)	Age cohort specific death rate (per 1000)	Deaths	Population (0–4)	Age cohort specific death rate (per 1000)
1956	15	125	120	1	129	78
1957	8	121	66	6	137	44
1958	13	139	94	1	142	71
1959	15	135	111	1	151	66
1960	12	145	83	2	162	12
1961	3	161	19	1	175	57
1962	2	196	10	2	182	11
1963	2	199	10	3	198	15
1964	2	221	9	6	212	28
1965	3	239	13	2	213	9
1966	2	231	9	2	222	9

five. Again, for Tunuru parish sample villages, no comparably sharp decrease in deaths is evident in the younger age group.

Turning to the number of recorded births (or, more properly, baptisms) per year for these groups (Figures 3.4 and 3.5 and Table 3.4), there has been a substantial increase for Manetai and Tunuru which is statistically significant (analysis of variance gives $p < .01$). The Eivo and Simeku evidently are the major contributors. There are two likely causes. First, the large postwar cohorts are beginning to reach marriageable age and are producing more children than the previous group, an "echo effect" to the early stages of the current population explosion. Second, it is likely that the postpartum taboos on sexual intercourse have become more honored in the breach than in the observance. Ogan et al. (1974) report a greatly shortened interval between pregnancies for some Nasioi and Nagovisi groups.

As a result, the yearly rates of increase for Manetai over this eleven-year period become substantially larger—from about 20 per thousand to approximately 50 per thousand in the final years—truly a remarkable rate of population explosion. The Tunuru parish populace began at a higher rate and expanded, but somewhat more slowly, attaining rates of increase of about 40 per thousand in the mid 1960s.

Birth Distribution Per Month for the Period 1956–1966

An unexpected finding which I came across by chance was that the distribution of births throughout the year in the Manetai Parish is not at all uniform. From what is known about the climate and growing season (or lack thereof) in Bougainville, and particularly in this region, one would not expect any

3.4   Manetai parish standardized monthly total births (1956–1966). (Standardized by finding the percentage difference of the observed monthly values from the expected values for each month)

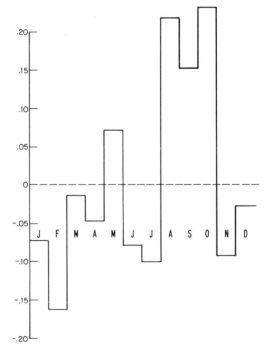

birth peak, or a specific time of year when most births occurred. However, as Table 3.5 and Figure 3.4 show, births in the Manetai area are concentrated in August, September, and October while there has been, during the last eleven years, a relative lack of births during February. I know of no comparable records for tropical nonurban groups. Birth peaks are known to exist for Eskimos (Weil 1971), where the differences in the seasons make a birth peak and trough reasonable findings. There also are different birth peaks for urban populations in Europe and the United States (Cowgill 1966). In Europe, more people are born during the spring, and particularly May, than during other seasons, while in the United States, the birth peak is in the late summer and autumn. The usual explanation for this difference is the social one that the European conception peak of August corresponds to the traditional vacation

3.5   Tunuru parish standardized monthly total births (1956–1966)

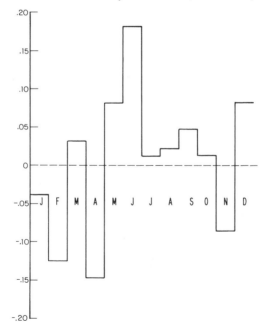

month there, while the American conception peak coincides with the period of Christmas and New Year's festivities.

Possibly there is a similar explanation for the Manetai area. Feasting is currently centered around Christmas, and it is rare to hear of a major pig feast occurring at other times of the year. Equally important, many absentee husbands return to their families during the Christmas season, if they are on the island, so that in this instance the birth peak may well be the result of Western influence. The analysis of variance results are significant at the .05 level.

Table 3.4
Parish Births, 1956–1966

Year	Number surviving births	Number born, but since dead	Total births	Population	Crude birth rate (per 1000)	Rate of increase (per 1000)
			MANETAI			
1956	19	15	34	635	54	18
1957	18	17	35	654	54	20
1958	41	11	52	685	76	45
1959	45	10	55	716	77	43
1960	31	14	45	744	60	37
1961	53	0	53	784	68	51
1962	50	0	50	824	61	49
1963	44	2	46	859 (877)	54	41
1964	65	2	67	922	73	69
1965	51	4	55	969	57	49
1966	60	0	60	1024	59	54
			TUNURU			
1956	25	3	28	757	33	28
1957	33	4	37	779	42	21
1958	36	0	36	810	45	39
1959	35	3	38	842	42	35
1960	36	0	36	868	41	29
1961	44	5	49	912	48	43
1962	41	3	44	942	44	29
1963	49	3	52	988 (1097)	50	44
1964	50	5	55	1030	49	36
1965	38	2	40	1066	36	32
1966	50	0	50	1136	44	40

Other possible causes for these results, particularly errors in the collection of the data, do not seem very likely. Births are reported to the missions promptly throughout this area, as baptisms should be performed within the first week of life, and in this regard the Bougainville parishioners are evidently model Catholics.

The records for Tunuru parish suggest a somewhat similar pattern, with a dearth of births in February, but the results are not quite significant. Possibly because many husbands in this area can more easily return home from jobs in Kieta and Arawa, the conception times are not so concentrated.

Retrospective Fertility and Patterns of Marriage (Tables 3.6 and 3.7).

By Oliver's and Ogan's accounts, it is the natural state of affairs for an adult on Bougainville to be married. However, while this was borne out in the survey for women, substantial numbers of men in these four language areas postponed marriage until their mid-twenties, although virtually everyone had been married at least once by the age of thirty-five.

Infertility is generally thought to be a problem of modern civilization. Other studies (e.g., Neel 1970) have claimed to show its virtual absence in traditional or technologically primitive societies. It is hard to establish its frequency

Table 3.5
Number of Births by Month (July 1956–June 1966)

	Manetai			Tunuru		
	Male	Female	Total	Male	Female	Total
January	34	30	64	51	45	96
February	20	33	53	39	41	80
March	34	34	68	57	46	103
April	31	33	64	47	36	83
May	34	40	74	50	58	108
June	34	28	62	62	53	115
July	37	25	62	60	41	101
August	34	50	84	44	58	102
September	28	49	77	41	61	102
October	44	41	85	51	50	101
November	37	24	61	50	39	89
December	29	38	67	48	60	108

Table 3.6
Female Retrospective Fertility: Number of Children Surviving Infancy by Current Age of Mother (January 1967) for Sampled Villages in Four Language Groups[a]

Number of children	Age of Mother											Total
	15–19	20–24	25–29	30–34	35–39	40–44	45–49	50–54	55–59	60–64	65+	
0	5	7	1	4	3	2	2	4	5	4	—	37
1	5	12**	9*	3	2	—	7*	3	5	4	2	52
2	2*	14	18	8	9	1	4	5	6*	2	5	74
3	—	9	22	15	11	6	6	3	4	1	3	80
4	—	1	8*	17	11	4	4	3	3	2	1	54
5	—	—	5	16	16	8*	5	5	3	3	3	64
6	—	—	—	12	10*	6*	4	5	3	—	—	40
7	—	—	—	—	7	12	4	1	1	1	—	26
8	—	—	—	—	—	2	—	—	—	—	—	2
9	—	—	—	—	—	—	—	—	—	—	—	0
10	—	—	—	1	—	—	—	—	—	—	1	2
Women	12	43	63	76	69	41	36	29	30	17	15	431
Children	9	71	172	294	286	212	126	96	81	41	50	1438
MEAN	.8	1.7	2.7	3.9	4.1	5.2	3.5	3.3	2.7	2.4	3.3	
TOTAL COHORT FERTILITY							3.5	3.3	2.7	2.4	3.3	
VARIANCE	.5	1.1	1.3	3.1	3.3	3.8	4.6	4.6	4.2	4.6	4.9	

[a] Women with questionable records are excluded, but their most likely assignations are indicated by asterisks.

Table 3.7
Male Retrospective Fertility: Number of Children Surviving Infancy by Current Age of Father (January 1967) for Sampled Villages in Four Language Groups[a]

Number of children						Age of Father						
	15–19	20–24	25–29	30–34	35–39	40–44	45–49	50–54	55–59	60–64	65+	Total
0	—	3	6*	10*	6	2	2	4*	—	3	2	38
1	—	1*	6*	10	5*	5	3	2*	2	1	3	38
2	—	2	14*	16*	11	7	2	7	4	1	5	69
3	—	—	9	18***	20	4	8	2	2	1	3	67
4	—	—	2	6*	18*	11	7	3	—	3	2	52
5	—	—	1	6	15*	12	11	8	2	—	3	58
6	—	—	—	—	11*	11	6*	6*	1	—	1	36
7	—	—	—	—	1	8**	5	4	1	1	1	21
8	—	—	—	—	—	1	4*	1	2	—	—	8
9	—	—	—	—	—	1	1	1	—	—	—	3
Men	—	6	38	66	87	62	49	38	14	10	20	390
Children	—	5	74	150	307	274	226	155	55	32	58	1336
MEAN	—	.8	1.9	2.3	3.5	4.4	4.6	4.1	3.9	3.2	2.9	
VARIANCE	—	.8	1.5	2.1	2.9	4.3	4.6	5.8	5.9	5.3	3.6	

a Men with questionable records are excluded, but their most likely assignations are indicated by asterisks.

in the Bougainville sample because of its confusion with the previously high rates of perinatal mortality. Certainly, marriages between healthy individuals between the ages of twenty-five and forty which have lasted longer than five years and which have produced no living children are rare. In the sample, there are two such Nasioi couples, one Torau, four Simeku, and four Eivo. Two men actively complained that their wives were barren (and attributed this to root medicines). But whether or not this is the case, infertility is not widespread.

The mean number of children surviving infancy for each five-year age category of women reflects the rising birthrate and lower infant mortality. Total cohort fertility (admittedly underestimated here) is clearly on the increase in the age classes passing through the child-bearing years under the currently favorable conditions, with an enormous jump in mean fertility occurring in the cohort just completing its fertility (age 40–44), and with more increases probably in store for the future.

Male retrospective fertility varies from the female pattern only in the delayed start; men marry, on the average, a few years later than women. The variance in production of offspring surviving infancy is not remarkably different in the two sexes, unlike the Xavante situation reported by Neel (Neel et al. 1964), where one or two powerful polygamous males contributed disproportionately to the gene pool of the offspring generation.

These data are all potentially important because evolution is ultimately dependent on the differential fertility of individuals as well as on their differential mortality. If early mortality rates are high, and if there is a wide spread among families in the number of children who survive to adulthood, then genetic changes can occur quite rapidly, if the differential mortality and fertility rates are based to some extent on genetic differences. This, in essence, is Fisher's Fundamental Theorem of Natural Selection, which he stated in more mathematical terms: "The rate of increase in fitness of any organism at any time is equal to its genetic variance in fitness at that time" (Fisher 1930).

We know almost nothing about the genetic component in the probability of survival of the average child or the reproductive performance of the average adult. In a recent article on the genetic aspects of reproductive fertility in a wide variety of societies, Neel and Schull (1972) suggest that potential opportunities for selection in mortality differentials are likely to be greater for agriculturalists than for hunter-gatherers and "primitives." That is, they suggest that mortality rates may have increased recently in traditional societies because of exposure to new diseases introduced by representatives of Western culture.

The Bougainville situation, at least, suggests the opposite. That is, contact with the West and Japan led, from almost the beginning, to a decrease in very high preadult mortality, while fertility levels are actually on the increase as well, as a result of the breakdown in postpartum taboos on sexual intercourse. In this sense, then, the extent of differential fertility among women appears to be greater today than in precontact times, so that the *possibilities* for selection acting through differential *fertility* (as opposed to mortality) are probably greater today than before. However, the genetic contribution to these differences in fertility, and all reproductive performance, seems to be quite small, by all accounts.

Inbreeding

I mentioned that the traditional preferred marriage in the Bougainville societies which have been studied by ethnographers was to marry a cross-cousin, real or classificatory. Sib exogamy, on the other hand, which is still very seldom violated, probably eliminates about 20 percent of one's potential marriage partners. In northeast Siwai, out of 646 recorded marriages, 14 were between members of different subdivisions of the same matrisib, and the sub-sibs involved are not reckoned to be very closely related. The Catholic Church has discouraged cousin marriage, so that what is often called preferential consanguinity is a less important factor in Bougainville groups today.

Inbreeding as a concept has many aspects, however, and the attempts to quantify it in statistical or probabilistic ways have been numerous (Wahlund 1928, Malécot 1948, Wright 1951). Inbreeding introduces the likelihood or possibility that a person will have a pair of genes which are identical because he has received two copies of one of his ancestors' genes, contributed to him through both his paternal and maternal lines. This probability of "identity by descent," as opposed to identity because of some other reason, is the coefficient of inbreeding, F. It can be calculated for an individual whose pedigree is known, or averaged for all individuals in a population. The analysis of pedigrees is ordinarily difficult because most people can only reconstruct their ancestry two generations on both sides. In Bougainville, I was fortunate to have access to genealogies from the northeast Siwai area which Oliver recorded in 1938–39 and which I used as an independent check and supplement to my own attempts at genealogical reconstruction in 1970. Even under these circumstances, unusual for a nonliterate society, some of the pedigrees were extended back only to the second generation, to the grandparental level. The pedigrees are from a group of seven neighboring villages in an area with a mean radius of 2.5 kilometers. They have been analyzed using the program COEF (MacLean 1969) which performs the calculation for the propositus:

$$F = \sum_{A} (1/2)^{n_1 + n_2 - 1},$$

where n_1 and n_2 are the number of generations between the common ancestor and the propositus through the father's line and the mother's line, respectively, and the summation is over all common ancestors. The resulting average inbreeding coefficient (F) is .008 for the current generation of this area, meaning that the likelihood that any particular individual will have two identical genes (be homozygous) at any given locus is .008 above what one would expect, given the frequencies of the different genes in the local population. Compared with inbreeding coefficients from other nonliterate societies where transportation is nonmotorized, this figure is neither startlingly high nor low. As in most societies where individuals are effectively limited to mates from their immediate neighborhoods (here taken to mean within a radius of 5 kilometers), a major restriction on matings is the one of geographical proximity.

As a simple exercise to illustrate the influence of the limits of the population size, I took the genealogies of the present generation sibships, and from the list generated random hypothetical marriages and then computed the resulting average inbreeding coefficient for the hypothetical offspring. For these, the result was $F = .006$, not appreciably different from the current generation results. The reason for, and the significance of, the exercise are simple. Because the population is small, and because migration in and out is relatively small, after a number of generations random mating within the restricted population will produce low levels of inbreeding, as most people will be related to each other to some degree. Given the gene frequencies in that small population, there will be a very small increased likelihood of identity of alleles over predicted (Hardy-Weinberg) expectations. More concerning inbreeding and its effects will appear in later sections. Here I simply have shown that the levels of inbreeding that are detectable in the parts of the sample with the best pedigrees are not at all extraordinary.

Migration

Current migration patterns within the villages in the sample evidently still follow approximately the same curve that characterized migration in the pre-war era. To judge by the distribution of distances that married individuals have moved from their birthplaces (Figure 3.6), the great majority of adults

3.6   Migration distances of married individuals from their birthplaces

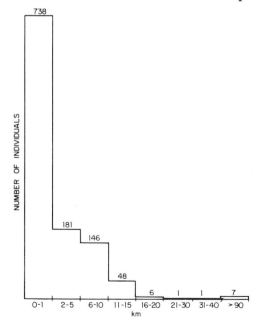

reside within a kilometer or two of their birthplaces. This steeply leptokurtic distribution of migration characterizes most sexually reproducing organisms, but the highly sedentary population shown by the steepness of the curve is extreme for humans. The curve for hunter-gatherers, such as the African Bushmen (Harpending and Jenkins 1973) is much less regular in its decline and has far fewer individuals in essentially the same locales as their places of birth. The Bougainville pattern is characteristic of people tied to the land, as tropical gardeners generally are. Another important feature of Figure 3.6 is that only a small trickle of people seem to move more than 10 kilometers away from their birthplaces. Over the sample, the average radius of a language group is approximately that distance. Most likely, this language pattern feature is a result, rather than a cause, of the sedentary migration pattern, which is more likely attributable to the subsistence dynamics of the group. That is, language differentiation probably was rapid and extreme *because of* the lack of movement and intercourse of the people, and later acted itself to restrict movements between people no longer speaking mutually intelligible tongues.

In any case, the extreme localization of individuals should make gene frequency variation over the entire area quite likely. The smaller the pool of available mates, the more likely will be fluctuations from one generation to the next in the frequency of the genes passed on. And the more such small populations there are, the more likely it is that the final populations will have diverse genetic patterns, or that the gene frequency variation among them will increase.

Prediction of Genetic Variation from Current Rates of Migration

It is possible to quantify the expected effects of such isolation on gene frequency fluctuations, and also to predict the patterns of similarities and differences between villages that should result if the villagers behave themselves and keep following the same patterns of marriage and migration over a number of generations. Essentially, because people will tend to marry others from their immediate or neighboring locales, there will be fluctuations in gene frequencies from one area to the next, which would tend to be smoothed out if the chances were equiprobable that individuals would take a mate from any distance. This is the effect of subdivision of a large population into many smaller groups. Even though individuals within the subdivisions might marry each other with no regard for kinship relations, over time individuals from each separate group would be more like one another than they are like those from other subdivisions. That is, they will share more genes in common than they will with other groups, and in a certain sense will be inbred, relative to the total population. Wright's (1943) formulation of the different sorts of "inbreeding" is as follows:

F_{IS} = the inbreeding of the individual relative to the local subdivision (recoverable from pedigrees),

F_{ST} = the inbreeding due to subdivision of the total sample into smaller units,

F_{IT} = total inbreeding, or the inbreeding coefficient of the individual relative to the total population,

and where

$$F_{IT} = F_{ST} + F_{IS}(1 - F_{ST}).$$

The coefficient of kinship is one quantity which has been used to express the effects migration and sample size (or genetic drift) should have on the gene frequencies of villages over a number of generations. It is defined as the probability that a gene taken at random from one village (i) will be identical because of common descent with a gene taken at random from another village (j) at the same gene locus. It is very closely related to Wright's inbreeding due to subdivision, F_{ST}. It estimates, in a sense, the variation and covariation of gene frequencies in subpopulations from an original (and unknowable) ancestral population. The formula for estimating the complete set of kinship coefficients among a set of village subpopulations used here is given by Imaizumi et al. (1971).[2] It is

$$\Phi^{(t)} = \sum_{k=1}^{t} (1 - s)^{2k} \mathbf{M}'^k \mathbf{U}^{(t-k)} \mathbf{M}^k. \qquad (1)$$

Here, k is the generation number, which goes from 1 to t,

s is the systematic pressure (here outside migration),

$\mathbf{M}$ is the initial migration matrix, and

2. This is a modified version of the method of Bodmer and Cavalli-Sforza (1968).

U is a diagonal matrix which has in its ith diagonal position $[1 - \phi_{ii}\,(k-1)]/(2\,N^e{}_i)$, where $N^e{}_i$ is the effective size of the ith subpopulation.

By way of general explanation, the exchange of mates among subpopulations is described by a matrix (M), which has in any i, j^{th} position the portion of the genes of subpopulation j which comes from subpopulation i during each generation.

Tables 3.8 and 3.9 present the raw data from which a modified matrix M

Table 3.8
Intermarriage Patterns of the 17 Villages Represented in the Sample[a]
(Current residences of individuals as rows; their birth places as columns)

Village[b]	1	Other Aita	Other Rotokas	2	3	4	5	6	West Eivo	7	North Simeku	8	9	10	West Simeku	Oraimi Nasioi	11	12	13
1	48	6	3																
2		1	34	8															
3			2	112	5	1	1	1				1							
4				1	23	6	1	1											
5					5	57		7											
6									34	16	1	6							
7									13		1	32	2		6				
8							1	2	2	1	9	37	4	4	1				
9												15	49	4	4				
10											3	2	1	89	2		4	4	1
11															1	9	16	4	
12											1						2	2	35
13															2	2		4	
17															1				1
18															1	1			1
14																			
15																			

[a]Composite group 16 is not included.
[b]Refer to Figure 2.3 and Table 2.1 for names & locations of numbered villages.

is calculated for this sample. This **M** matrix is then taken to be constant from one generation to the next, with its proportions of exchange fixed. The contribution to the genetic variance of the new generation, then, is determined by multiplying the i, j^{th} element by the j, i^{th} element of the previous generation. The migrants are assumed to be a random sample of the inhabitants of their subpopulation of origin. The new population undergoes simulated genetic drift each generation by binomial sampling of the parental gametes, which is what the **U** matrix accomplishes. The larger the population, the smaller will be the contribution to the variance by drift.

Other N. Nasioi	South Nasioi	Koromira Nasioi	17	Other Torau	18	Central Siwai	Other NE Siwai	14	15	Terei	Nagovisi	Nissan	New Ireland	New Guinea	Manus	N
																57
																43
																123
																32
														1		70
																57
																54
														1		62
																72
												1				107
8											1					39
30	1													1		72
16	2	1				1										42
5	2	35	3	1										1	1	50
1			15	101									1			121
						12	12	32		1						57
						6	11		47	6						70
																Total = 1128

Table 3.9
Intermarriage Frequencies of the Linguistic Groups Represented in the Sample
(Current residences of married individuals as rows, their birthplaces as columns)[a]

Village	Aita	Rotokas	Eivo	Simeku	Nasioi	Uruava	Torau	Siwai	Buin	Nagovisi	Nissan	New Ireland	New Guinea	Manus	N
Aita	.9474	.0526													57
Rotokas	.0233	.9767													43
Eivo		.0060	.8507	.1403											335
Simeku			.0373	.9047	.0498								.0030		241
Nasioi				.0196	.9542	.0065		.0065		.0065	.0041		.0041		153
Uruava				.0200	.1600	.7000	.0800						.0065		50
Torau							.9587					.0083	.0200	.0200	121
Siwai				.0165	.0165			.9449	.0551						127
														Total = 1127	

[a]The data are taken only from the 17 villages covered in this study.

Fixation, or loss of all but one allele, which would otherwise inevitably occur in such a closed system, is prevented by linear systematic pressure (here $1 - s$), which is assumed to be immigration from an effectively infinite or constant outside population. The assumption that long-range migration, rather than selection, mutation, and so on, is the agent of systematic pressure keeping gene frequencies at some intermediate value is necessary so that all genes may be described by the same model. Under selection, for example, the systematic pressure would be different for each allele. In field studies systematic pressure is not readily ascertainable. Some empirical experience indicates that variation in s does not have a large effect on gene frequency variation within a population if the local inbreeding within the population is low. Only as the migration matrix approaches the identity matrix should systematic pressure become important in predicting local variation (see Harpending and Jenkins 1973).

Here is an example of the mechanics involved in the computation of a single relationship coefficient after one generation of migration and drift. If the effective sample size of the ith subpopulation is 50, and .90 of the current married residents were born in the village, and if we estimate the systematic pressure to be .05, then by equation (1),

$$\phi_{ii}^{(1)} = (1 - .05)^2\, (.90)\, (1/100)\, (.90) = .0073.$$

This represents the estimate of the variance of the new gene frequency of this subsample due to the fact that only a fraction (.90) of the previous generation will mate with itself to produce the next, and that the size of the sample itself will influence the variance.

For the second generation mating in this pattern the resulting variance contribution will be less, as only .81 of the genes from the original population remain to undergo random mating, so that the additional contribution to the variance of the new gene frequency from the original frequency will be

$$(1 - .05)^4\, (.81)\, (.01)\, (.81) = .0052, \text{ so that } \phi_{ii}^{(2)} = .0125.$$

In this way, the variances around the initial gene frequency of the population should slowly increase, albeit at a diminishing rate, until the differences between two succeeding matrices will be infinitesimal, and the matrices will be said to be in equilibrium. This has little meaning in reality, where it is highly improbable that the same mating pattern will persist for 200 successive generations, so in this study I have stopped after 50 generations (which pushes reality as it is).

So the matrix Φ or some statistic derived from it, seems to be a natural measure of the amount of genetic sampling or drift to which a population has been subject, if there is some reliable estimation of the amount of gene exchange which has occurred through the last 1,000 years (highly unlikely). Two natural measures are the average relationship (or random kinship) of all groups,

$$\overline{\phi_{ij}} = \sum_{i,j} w_i w_j \phi_{ij}$$

and the average inbreeding coefficient, $F_{ST} = \sum w_i \phi_{ii}$.[3]

The kinship matrix Φ resulting from the analysis of the Bougainville data for the 15 villages on the eastern coast (the Siwai villages, being isolated from the rest were not appropriate to include) is given in Table 3.10. These statistics, however, estimate variation or movement from an original (and unknowable) condition, whereas what will be dealt with throughout the rest of the book are observed biological variations in *current* populations around their *current* averages. Harpending and Jenkins (1973, Appendix 1) show that a more useful statistic in this case is what they call the relationship coefficient (r), a measure of current variances and covariances around the current mean which can be estimated from observed gene frequency data. The opposite passage, the estimation of inbreeding from current gene variation, is not really possible. The expectation of the r statistic from the Φ matrix is:

$$E(r_{ij}) = \frac{\phi_{ij} + \overline{\phi} - \overline{\phi_i} - \overline{\phi_j}}{1 - \overline{\phi}}$$

The resulting matrix of r values is given in Table 3.10.

The natural summary statistic of an **R** matrix is $\sum_i w_i r_{ii}$, or the variance of a gene frequency standardized by dividing by $\overline{pq}$, written $\sigma^2/\overline{pq}$. This is equivalent to what is elsewhere called Wahlund's f (e.g., Cavalli 1973). Wahlund's f coefficients from completely different sets of populations will be heavily influenced by at least two extraneous factors, the sample size and the way the sample

3. w_i is the effective size of population i.

Table 3.10

Φ and R matrices predicted from current rates of migration among 15 Bougainville villages[a]

(each figure × 10^{-4})

Elements of the Φ matrix appear in the upper right; elements of the R matrix in the lower left triangle.

Villages	–	1	2	3	4	5	6	7	8	9	10	11	12	13	17	18
1	487	530	50	0	0	0	0	0	0	0	0	0	0	0	0	0
2	9	529	560	20	0	0	0	0	0	0	0	0	0	0	0	0
3	−44	−24	316	360	60	10	10	0	0	0	0	0	0	0	0	0
4	−39	−33	19	446	472	150	20	40	0	0	0	0	0	0	0	0
5	−44	−40	−41	107	480	520	40	70	0	0	0	0	0	0	0	0
6	−46	−42	−42	−24	−6	271	320	300	100	0	0	0	0	0	0	0
7	−47	−43	−50	−10	18	248	261	310	60	10	10	0	0	0	0	0
8	−34	−30	−33	−32	−34	62	16	173	388	140	30	0	0	0	0	0
9	−39	−36	−42	−38	−41	−31	−36	108	13	430	50	0	0	0	10	0
10	−39	−35	−42	−37	−42	−38	−7	−36	−37	234	437	70	10	20	0	10
11	−36	−35	−39	−37	−43	−44	−37	−30	−35	35	270	308	150	70	0	0
12	−38	−32	−41	−34	−40	−41	−42	−25	−35	−27	113	470	606	100	10	10
13	−37	−34	−40	−36	−41	−43	−44	−31	−28	−20	31	65	340	640	70	10
17	−37	−33	−40	−36	−36	−43	−44	−31	−34	−34	−32	−26	37	293	330	110
18	−45	−41	−48	−43	−48	−50	−51	−38	−29	−40	−36	−40	32	71	238	290
	1	2	3	4	5	6	7	8	9	10	11	12	13	17	18	–

[a] Φ = Kinship coefficients; R = "Relationship" coefficients.

sizes are defined. Especially because of these problems I am now somewhat dubious about the value of comparisons of values of f from different populations (see, for example, Cavalli-Sforza 1969, and also Friedlaender 1971b). The major cause of the differences in such comparisons is most likely the different sample sizes in the different groups, with small sample sizes producing large values for ϕ and r. In the present case there are so many uncertainties and obvious false assumptions (for example, the uniform systematic pressure, the stable effective size of the population, the fleeting reality of the values for the **M** matrix) that the results are only general "ball-park" estimates for the pattern of variation expected, as well as the degree of variation.

In a previous publication (Friedlaender 1971a, p. 22), I calculated estimates of the average coefficient of inbreeding, utilizing not only the information on migration among the 15 villages, but also among neighboring, but unsurveyed villages. The resulting value for inbreeding of villages relative to the total sample array, $F_{ST} = .04916$, may appear quite high, especially in comparison with the pedigree estimate of inbreeding from northeast Siwai just given as $F = .008$. However, the two estimates are of quite different parameters. The first F_{ST}, is a measure of genetic heterogeneity among villages over the Bougainville sample which would be the end result of current migration rates among them after many generations. The second, F, is the calculated average inbreeding coefficient, or identity by descent, among villagers in a small section of Siwai, where pedigrees are adequately known for only the last two generations. It is not unreasonably small by comparison.

Because the focus of this entire survey is not only the amount of variation among these villages but also the pattern of variation, I subjected the **R** matrix of predicted genetic variances and covariances from current rates of migration to a principal components analysis, which yields a description of variation in a small number of dimensions.

The new coordinate axes of the components analysis reflect the major axes of variation among the 15 villages included in the migration analysis. The plots of the villages on the first three coordinates should fairly accurately reflect the pairwise distances between the individual groups, as depicted in Figure 3.7, although all of the 14 principal components would have to be plotted in order to have a perfectly accurate representation of the distances. As it is, they serve to give a visual interpretation to the distance matrix which will be utilized at a later stage of the book (see Chapter 8).

Note that, generally speaking, the villages are more separated in the northern sector of the sample and that villages 1 and 2 are so widely separated as to have no close relationship with any of the rest. Also, although they have quite distinctive languages, the "Melanesian" group (villages 17 and 18) are interbreeding extensively with their neighbors so that their relationships predicted from current rates of migration alone are closest to the Simeku.

All this may seem like gimmickry at this point, given the many qualified statements, but this is the simplest and most efficient approach currently available for comparing variation within a number of different biological and demographic variables.

3.7   Array of 15 sample villages (excluding Siwai) predicted from current rates of migration. First three axes

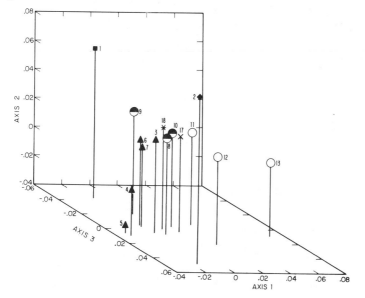

20. Collecting Genealogies

21. Anthropometry

Hypotheses

My primary assumption is that the villages, themselves conglomerations of older hamlets in close geographical proximity, were small enough and isolated enough in traditional and precontact times to develop very considerable heterogeneity in gene frequencies and hence in genetically determined variation. The results of the analysis of current migration patterns, which predict the similarities of the samples for some future time given a constancy of current migration rates, serve as an initial indicator of genetic relationships of the different groups, but only insofar as the current rates reflect marriage exchanges in the past. Because there are no marked social strata, I also expect gene frequencies to vary with increasing geographic distance, with those villages closest together

22. Blood sampling

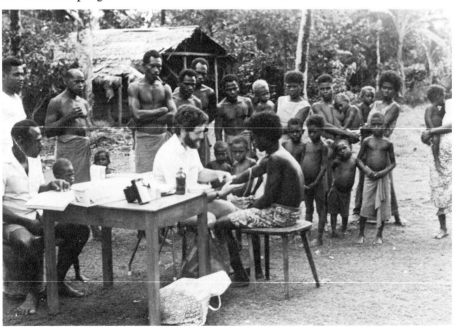

23. Assistants Michael Mandaku (Rorovana) and Ignatius Bakamori (Boira)

exhibiting the greatest similarities. This should be conditioned and qualified by the language barriers in the area, which presumably reflect old and stronger migration restrictions than are currently in force.

If physical variation on the island follows a distinguishable pattern, it would almost certainly be related to the migration predictions, to geographic proximity of different villages, and to language groupings, with their implications for historical ties as well as marriage patterns. The language groups might form distinguishable clusters phenotypically as most marriages are made within the linguistic unit. Border villages [such as Uruto (7), Bairima (13) and Arawa (17)], where most bilingual individuals in the study reside, should be intermediate. However, as the rate of marriage across linguistic boundaries is, and was, considerable, it will be interesting to note those characteristics which reveal differences corresponding to the linguistic groupings and those which show no variation at all, or which vary in some other fashion.

Another consideration from the demographic information is the differential density of the population aside from the linguistic and geographic factors. The southern portion of the area covered in the survey is relatively densely populated (see Figure 2.2). Moving northward through the Simeku, Eivo, and southern Rotokas areas, the population becomes increasingly sparse and oriented along a single footpath. Further to the north the number of villages increases, and their arrangement becomes less linear. The groups in the sparsely populated strip of land at the center of the island might well be biologically intermediate in all characteristics between the larger populations to their immediate north and south. This would also fit well with the linguistic relationships of the groups. Moreover, there is the additional possibility that in this sparsely populated area, a number of alleles might be completely lost owing to random genetic drift. That is, this area may act as an effective bottleneck to gene flow.

This basic pattern may be secondarily affected by influences emanating from centers of westernization, in this case principally from Kieta to Tunuru Mission and inland thence to the central mountains and the mining area. Changes related to improvements in disease control or in nutrition should be manifest in differences between older and younger generations or between villages within, and on the periphery of, effective European influence. Since older historic epidemics and differential disease patterns within the island are poorly documented, it would be totally impossible to attribute any biological variation within Bougainville to disease effects in a positive way.

Altitudinal gradients in physical variation are, of course, possible. It is conceivable that the coastal dwellers, particularly the Melanesian-speaking Torau and Uruava, have been subjected to disease pressure from malaria more uniformly and intensely than some of the hill people, such as the Aita or Rotokas, but this would seem unlikely, since even these groups, or their representatives in this sample, unlike the New Guinea highlanders, are far from being malaria-free. Other environmental determinants, which might be used to explain north–south or altitudinal gradients in physical variation, would be unlikely, owing to the uniform climate of the island.

Differences in soil composition undoubtedly account in some way for biological differentiation within the populations. As an example, goiter, with a high incidence rate over most of Bougainville, is particularly severe in Atamo, Uruto, and Karnavito villages, situated on the leached volcanic soils of the Alakabau River drainage, as mentioned in Chapter 2. Genetic correlates, however, are not immediately obvious. The ability to taste phenylthiocarbamide, reportedly more frequent in populations in iodine-deficient environs, is not any more common in this area than in the rest of the Bougainville sample (see Appendix).

Now that the geographic, linguistic, epidemeologic, and demographic relations of the villages have been delineated, the following chapters will describe the extent to which current human biological variation can be accounted for within this framework.

4 Blood Genetics

Since World War II the increasing number of known and frequently occurring blood variants (or polymorphisms) has become the principal focus of human population studies because their mode of inheritance is clearly understood in most cases, and because environmental factors evidently cannot modify the majority of them through the life span. In a period of activity comparable to the late nineteenth century, when physical anthropologists were excited by the systematic method of description which measuring parts of the body seemed to afford, the energies of physical anthropologists and human geneticists were spent on typing the blood of exotic and different groups of humans and other primates, the rationale being that these blood variants reflected changes in the real genes, the important elements determining variation in evolution.

There was the fuzzy notion, as well, that variation in blood antigens and proteins somehow was not open to the effects of natural selection, and would therefore reflect population and race history more accurately than obvious variation in body shape, size, and color. As I mentioned in the Introduction, this controversy still persists in some forms, but no one with any familiarity with the problem will deny that it is highly probable that natural selection affects gene frequencies which control polymorphic variation in at least some blood proteins.

In actively looking for repetitive patterns of genetic variation, there is the possibility that I am introducing a bias in my approach. Selectionists rightly argue that each genetic locus may be subject to its own unique history of evolutionary events and that pooling the results over many unrelated loci is contrary to evolutionary reality. There seems to be little argument that the human ABO and Rh blood groups, for example, are subject to considerable selection pressures, although the exact nature of these pressures has yet to be understood on even the most rudimentary level.

My position here is that this survey can do little to elucidate the nature of selection for particular systems by an examination of the separate gene frequencies in the different villages by themselves. Over such a small area, it is reasonable to assume a constancy of selection pressures for each system.[1] In-

1. If readers are interested in particular aspects of the blood genetic distributions, they are referred to earlier publications (Friedlaender 1969, 1971 a and b, 1975, Friedlaender et al. 1971). Consult the Appendix for a description of the methodology.

stead, my object is to establish the general nature of the variation in seven different genetic systems over the village samples, what they reveal concerning the overall genetic relationships of the different groups, and what they show about patterns of heterozygosity within groups (see Plate 22).[2]

Results

1. The Blood Groups

The ABO *system.* The ABO results for the 18 groups in this survey are presented in Table 4.1.

The fit of the estimated allele frequencies to the observed phenotype numbers under assumed random mating is very good and nowhere is the difference statistically significant. Subgroup A_2 was completely lacking, as in other populations from this part of the world.[3] The variation of phenotype and gene frequencies in this series is remarkable, I^A ranging from .15 to .48, I^O from .41 to .77 and I^B from 0.0 to .15. But most strikingly, the I^B allele is almost totally absent over the area from Okowopaia (2) through Sieronji (11). Two type B individuals, one from Boira (9) and one from Korpei (10), were born outside this area, one coming from New Guinea and one from Pomaua (12), in Nasioi territory.

This area lacking the I^B allele covers the entire Eivo (3–7) and Simeku (8–10) sample territory as well as the villages of Sieronji (11) to the south and Okowopaia (2) to the north. Recalling the demographic situation outlined in Chapter 3, this allele absence is in an area of low population density, but not complete population isolation. Given this probable loss of I^B, it is surprising that, while two language samples are lacking in the allele, so are two villages just beyond their boundaries. Both of these villages have recently migrated from other areas—Sieronji (11) from the mountainous Nasioi area directly west of their present locale, and the Okowopaians (2) from the very mountainous area inland from their new location. Both populations possibly

2. For a description of the nature of the individual polymorphisms, see Giblett (1969).
3. With the exception of the single celebrated family belonging to the Agarabi linguistic group of the Eastern Highlands (Simmons et al. 1961), no A_2 individuals have been found in Melanesian populations.

could have lost the I^B allele independently. Both are rather small groups, and the loss of this allele would have meant the exclusion (by chance or otherwise) of no more than five or six people.

The Eivo (3–7), linguistically a northern Papuan group, at one time must have moved down to the fertile region south of the dividing swamp and narrow waist of the island, which, with the exception of the Eivo, neatly divides the northern and southern Papuan-speakers on Bougainville. The Eivo may have lost the I^B allele in the process of founding this new settlement, or, with the Simeku and others, may simply have lost it over generations in this marginal underpopulated area. This matter will be discussed at greater length below.

The MNSs *System* (Table 4.2). As no crossover has yet been detected between the MN and Ss loci, it has become accepted procedure to present the results together—of L^{MS}/L^{NS}, and the like. For some populations—the Eskimos,

Table 4.1
The ABO System

Group	Observed phenotypes				Allele frequencies			Std. dev. of estimate			N
	A	O	B	AB	I^A	I^O	I^B	I^A	I^O	I^B	
1.	74	19	12	11	.485	.410	.105	.039	.039	.021	116
2.	47	51	0	0	.279	.721	.000	.280	.028	.000	98
3.	75	105	0	0	.236	.764	.000	.020	.020	.000	180
4.	22	31	0	0	.235	.765	.000	.040	.040	.000	53
5.	72	29	0	0	.464	.536	.000	.031	.031	.000	101
6.	66	51	0	0	.340	.660	.000	.051	.051	.000	117
7.	39	54	0	0	.238	.762	.000	.028	.028	.000	93
8.	58	68	1	0	.263	.733	.004	.030	.030	.004	127
9.	39	57	0	0	.229	.771	.000	.033	.033	.000	96
10.	85	119	1	1	.237	.759	.004	.022	.022	.003	206
11.	27	35	0	0	.249	.751	.000	.031	.031	.000	62
12.	62	47	5	1	.329	.645	.026	.035	.035	.010	115
13.	24	25	9	2	.249	.655	.097	.043	.047	.028	60
14.	20	50	14	5	.151	.737	.112	.028	.034	.024	89
15.	59	40	12	3	.211	.721	.068	.029	.031	.017	114
16.	10	13	2	1	.321	.620	.059	.072	.074	.033	26
17.	25	52	27	5	.149	.692	.140	.025	.033	.026	109
18.	117	65	39	25	.349	.511	.040	.024	.025	.016	246

for example—this close linkage is reflected in the unexpected frequencies of the binomial combinations. Therefore, the different MN and Ss alleles are given here in combined fashion (Table 4.2).

The frequency of L^{NS} increases in a cline from north to south. A marked break occurs in the Simeku-Nasioi transition zone, between samples 10 and 11. North of the border between the two, all the groups but one have L^{NS} frequencies below .05, and that one is Boira (9), the Simeku village which was traditionally closest to the Nasioi. The Nasioi, Torau, Siwai, and Uruava all have higher frequencies of L^{NS}. The Siwai have the highest L^{NS} frequencies of all, and the lowest L^{MS}, all under .10.

Heterozygote excesses in different MN mating combinations have not been analyzed, but the "goodness of fit" values for the distribution of phenotypes in the group samples are quite good. The only significant values are for villages with homozygote, not heterozygote, excess.

Table 4.2
The MNSs System

Group	Observed phenotypes						Allele frequencies				
	NsNs	NNS	MsNs	MNS	MsMs	MMS	L^{NS}	L^{NS}	L^{MS}	L^{MS}	N
1.	76	0	36	0	4	0	.810	.000	.190	.000	116
2.	64	2	30	1	1	0	.819	.013	.166	.002	98
3.	107	1	62	0	10	0	.768	.003	.229	.000	180
4.	29	4	14	1	5	0	.715	.048	.236	.000	53
5.	69	3	26	0	3	0	.827	.015	.158	.000	101
6.	68	4	37	3	2	3	.767	.020	.189	.024	117
7.	65	4	24	0	0	0	.839	.032	.129	.000	93
8.	76	6	30	2	13	0	.740	.032	.228	.000	127
9.	60	14	21	0	1	0	.804	.076	.120	.000	96
10.	148	6	43	1	8	0	.837	.017	.146	.000	206
11.	31	5	19	4	3	0	.706	.060	.219	.015	62
12.	47	19	34	6	9	0	.632	.116	.252	.000	115
13.	36	13	7	3	1	0	.766	.134	.100	.000	60
14.	26	15	8	1	0	0	.741	.175	.084	.000	50
15.	80	21	9	2	0	0	.830	.113	.057	.000	112
16.	14	8	3	0	1	0	.711	.193	.096	.000	26
17.	43	21	34	5	6	0	.639	.127	.234	.000	109
18.	87	24	99	15	20	1	.605	.078	.310	.007	246

The Rh *System* (Table 4.3). The Rh allele frequencies are the most consistent and stable of the reported polymorphic systems in this series. The range in R^1 frequencies over all 19 groups is a mere .14, and no pattern in the variation seems to be discernible. There is no cline.

The Rh *Factor* D^u. One individual, from Okowopaia (2), without known relatives, is listed as having the D^u+ variant. One would have to assume this was a typing error unless the sample were retaken and retested.

Kell and Duffy. The allele *k* is fixed over the entire series, as it seems to be in the rest of Oceania. The Fy^a allele of the Duffy system evidently also has a frequency of 1.00. These systems have therefore not been included in the

Table 4.3
The Rh System

Group	Observed phenotypes				Allele frequencies			Std. dev. of estimate			N
	R^1R^1	R^1R^2	R^2R^2	R^1R^0	R^1	R^2	R^0	R^1	R^2	R^0	
1.	182	33	1	0	.849	.151	.000	.020	.020	.000	116
2.	73	23	2	0	.862	.138	.000	.025	.025	.000	98
3.	145	25	8	2	.881	.114	.006	.017	.017	.004	180
4.	47	5	1	0	.934	.066	.000	.023	.023	.000	53
5.[a]	83	16	1	0	.906	.089	.000	.021	.020	.000	101
6.	92	24	1	0	.889	.111	.000	.020	.020	.000	117
7.	71	21	1	0	.876	.124	.000	.021	.021	.000	93
8.	114	13	0	0	.949	.051	.000	.013	.013	.000	127
9.	93	3	0	0	.984	.016	.000	.009	.009	.000	96
10.	165	37	3	1	.893	.104	.002	.014	.014	.002	206
11.	57	4	1	0	.952	.048	.000	.019	.019	.000	62
12.	96	18	0	1	.917	.078	.004	.018	.018	.000	115
13.	52	8	0	0	.933	.067	.000	.022	.022	.000	60
14.	82	6	1	0	.955	.045	.000	.015	.015	.000	89
15.	98	13	3	0	.917	.083	.000	.018	.018	.000	114
16.	24	2	0	0	.962	.038	.000	.027	.027	.000	26
17.	94	15	0	0	.931	.069	.000	.017	.017	.000	109
18.	177	57	3	9	.854	.128	.018	.015	.014	.006	246

[a]Village 5 has one R^1R^z tested individual, for a R^z gene frequency of .005, and a standard deviation of .005 for the estimate.

statistical analyses of these 18 groups, although the findings are interesting in broader comparisons.

2. *Serum Proteins and Red Cell Enzymes*

Transferrins. The Tf^D_1 variant was found in only two blood samples from Nupatoro (1), one from Korpei (10), one from Rorovana (18), and two from Arawa (17). Because of the ideomorphic character of this allele, the transferrin results were not included in the multivariate analyses which follow. The results will be compared with the findings of other surveys, however.

Haptoglobins (Tables 4.4 and 4.5). The Hp^1 and Hp^2 alleles vary over the Bougainville sample area in a way similar to that of the I^B allele. Groups 1 and 2, and 17 and 18 are the highest in Hp^2, while the poorly populated central area is a depressed one in terms of Hp^2. Again, variations in frequencies

Table 4.4
The Hp System

Group	Observed phenotypes			Allele frequency		Std. dev. of allele est.	N
	1-1	2-1	2-2	Hp^1	Hp^2		
1.	22	65	24	.491	.509	.034	111
2.	13	51	25	.433	.567	.037	89
3.	75	81	19	.660	.340	.025	176
4.	35	16	1	.827	.173	.037	52
5.	61	35	6	.770	.230	.030	101
6.	59	42	13	.702	.298	.030	114
7.	39	31	7	.708	.292	.037	77
8.	64	52	5	.744	.256	.028	121
9.	20	21	3	.693	.307	.049	44
10.	105	57	4	.804	.196	.022	166
11.	37	23	1	.795	.205	.037	61
12.	54	49	12	.683	.317	.031	115
13.	30	24	4	.724	.276	.041	58
14.	31	34	9	.649	.351	.039	74
15.	40	58	13	.622	.378	.033	111
16.	15	9	2	.750	.250	.060	26
17.	28	54	23	.524	.476	.034	105
18.	62	108	63	.498	.502	.023	233

Table 4.5
Hypohaptoglobinemia in the Sample of Bougainville Populations

Village	Number of hypohapto-globinemic individuals
1.	0
2.	2
3.	1
4.	1
5.	1
6.	1
7.	0
8.	2
9.	0
10.	1
11.	0
12.	0
13.	1
14.	1
15.	0
16.	0
17.	0
18.	1
Total	12

are very large, and reach the upper extremes of Hp^1 values reported from any worldwide sampling so far. Hypohaptoglobinemia (Hp 0) is present in low levels throughout this series (see Table 4.5). This phenomenon is usually attributed to malarial infection.

Red Cell Acid Phosphatase (Table 4.6).[4] The sample frequencies cluster into two distinct groups, north and south, with respect to this polymorphism; samples 1–10 and 11–18. This is similar to the L^{NS} pattern. The Nasioi have the extreme frequencies in the southern group in this instance, not the Melanesian-speakers. The range in P^a frequencies in Bougainville villages from .02 (for group 12) to .35 (group 9) is greater than the range of all previously

4. The acid phosphatase starch-gel electrophoresis runs were performed on 1165 samples of the present Bougainville series by Dr. L. Y. C. Lai at the University of New South Wales.

reported total population frequencies. Groups 9 and 12 are only about six miles apart at the same altitude, so their great divergence in frequencies is most likely due to the historically demographically fragmented situation on Bougainville.

Gm (Table 4.7). The Gm results (polymorphisms of the IgG chains) vary a great deal over the 18 samples, but particularly so for the two northernmost villages. Okowopaia (2) has an extremely high Gm^1 frequency (.438) and no $Gm^{1,2}$, while Nupatoro (1) has the highest value of $Gm^{1,2}$ (.121) and a low frequency of Gm^1 (.076). Villages 9 and 11 are fixed for $Gm^{1,3,5,13,14}$. This allele fluctuates from 1.00 to .56, a .44 range of variation for this allele frequency.

Inv (Table 4.8). The Inv frequencies (polymorphisms of the IgG chains) have the most striking distribution in this series of any hematological materials.

Table 4.6
The PHs System

Group	Observed phenotypes			Allele frequencies		Std. dev. of allele est.	N
	A	AB	B	p^a	p^b		
1.	7	38	47	.283	.717	.033	92
2.	1	23	34	.216	.784	.038	58
3.	6	21	51	.212	.788	.033	78
4.	6	29	29	.320	.680	.041	64
5.	4	53	59	.263	.737	.029	116
6.	5	29	36	.279	.721	.038	70
7.	5	15	22	.298	.702	.049	42
8.	2	28	53	.193	.807	.031	83
9.	1	8	5	.357	.643	.091	14
10.	15	58	81	.286	.714	.026	154
11.	0	6	30	.083	.917	.033	36
12.	0	1	26	.019	.981	.018	27
13.	0	1	4	.100	.900	.095	5
14.	0	13	66	.082	.918	.022	79
15.	2	30	62	.181	.819	.028	94
16.	0	6	19	.120	.880	.046	25
17.	1	2	11	.143	.857	.066	14
18.	1	18	56	.133	.867	.028	75

Table 4.7
The Gm System

Group	Observed phenotypes				Allele frequencies			Std. dev. of estimate			N
	Gm(1)	Gm(1, 2)	Gm(1, 2, 3, 5, 13, 14)	Gm(1, 3, 5, 13, 14)	Gm^1	$Gm^{1,2}$	$Gm^{1,3,5,13,14}$	Gm^1	$Gm^{1,2}$	$Gm^{1,3,5,13,14}$	
1.	1	2	21	75	.076	.121	.803	.041	.024	.044	99
2.	18	0	0	76	.438	.000	.562	.046	.000	.046	94
3.	0	1	2	187	.046	.008	.946	.035	.005	.035	190
4.	1	0	3	66	.108	.021	.871	.057	.012	.058	70
5.	1	0	3	124	.082	.012	.906	.043	.007	.043	128
6.	0	1	6	117	.000	.032	.968	.000	.011	.011	124
7.	3	0	1	86	.179	.006	.815	.051	.006	.053	90
8.	4	0	5	124	.163	.019	.818	.041	.008	.042	133
9.	0	0	0	93	.000	.000	1.000	.000	.000	.000	93
10.	8	0	0	173	.209	.000	.791	.000	.036	.036	181
11.	0	0	0	62	.000	.000	1.000	.000	.000	.000	62
12.	4	0	2	110	.181	.009	.811	.034	.005	.811	116
13.	0	2	4	53	.000	.068	.932	.000	.023	.023	59
14.	0	2	13	73	.000	.097	.903	.000	.097	.903	88
15.	0	0	13	99	.000	.058	.942	.000	.015	.015	112
16.	1	0	3	22	.164	.058	.779	.089	.033	.092	26
17.	1	0	6	101	.082	.028	.890	.044	.011	.045	108
18.	1	0	14	228	.050	.029	.921	.029	.008	.030	243
19.	3	2	9	149	.139	.034	.827	.037	.010	.037	163

The frequency of Inv^1 ranges from .826 at the northern extreme down to .326 for the Melanesian-speaking village of Arawa (17). The difference of .500 between the two frequencies is greater than fluctuations in any other system, and the regularity or the change is only approximated by the L^{NS} results, and then only over a much smaller range of values. It is remarkable that over the transitional area from the Simeku (8–10) through the North Nasioi (11–13), the range of frequency variation is a mere 7 percentage points. Recall, too, that outbreeding frequencies through this area are all relatively high. The Eivo, directly to the north, range from .707 to .637 in Inv^1 frequency with village 7, classified as Eivo but bilingual in Eivo and Simeku, falling midway between the other Eivo and the Simeku in Inv frequencies. The extreme northern villages (1 and 2) are highest in Inv^1. The Siwai fall in the same frequency range as the Nasioi, their linguistic cousins. The Melanesian-speakers, historically the

Table 4.8
The Inv System

Group	Observed phenotypes		Allele frequencies		Std. dev. of allele est.	N
	$Inv(1^+)$	$Inv(1^-)$	Inv^{1+}	Inv^{1-}		
1.	96	3	.826	.174	.049	99
2.	87	7	.727	.273	.050	94
3.	165	25	.637	.363	.034	190
4.	64	6	.707	.293	.057	70
5.	116	12	.694	.306	.042	128
6.	109	15	.652	.348	.042	124
7.	71	19	.541	.459	.047	90
8.	97	36	.480	.520	.037	133
9.	55	29	.412	.588	.044	84
10.	119	62	.415	.585	.030	181
11.	42	20	.432	.568	.052	62
12.	85	31	.483	.517	.040	64
13.	43	16	.479	.521	.056	59
14.	63	25	.467	.533	.045	88
15.	69	43	.380	.620	.037	112
16.	18	7	.529	.471	.085	25
17.	59	49	.326	.674	.036	108
18.	149	94	.378	.622	.025	243

most southern group, have the lowest frequencies of the villages in this series. The variation over this small section of Bougainville is remarkable. In fact, group 1 (Nupatoro) has a frequency higher than any other reported sample in the world (Gallango and Arends 1967).

3. Patterns in Allele Frequency Shifts Within This Series

There are some recurring patterns in the serological changes over this series. Frequencies seem to vary in one of three ways.

(A) Some frequencies vary in regular "clinal" fashion from one geographic extreme to the other. Allele frequencies of the Inv and PHs systems, as well as the L^{NS} allele of the MNSs system are examples. In these three systems, the major shift in frequencies occurs in the transitional Simeku (8–10) area. The Inv cline is steepest at the Eivo–Simeku boundary, while the other two are steepest at the Simeku–Nasioi boundary.

(B) A second pattern may be described as a "U-shaped fluctuation," where the most diverse groups, in terms of geographic distance and historical relations, have the most similar frequencies. This is the characteristic pattern of the Hp alleles and the I^B allele of the ABO system. I^B is effectively lost over the area from sample 2 to 11, while it has frequencies ranging up to .15 in the populations on either side. This pattern is corroborated in the other Bougainville studies discussed below. The Hp^1 and Hp^2 alleles vary in the same way, except that group 2 belongs to the extreme group rather than to the central section of the "U."

(C) A third group fluctuates in "random," that is, unintelligible, fashion which cannot be positively attributed to the effects of any directional agent of change. These include the Rh and MNSs alleles (except for L^{NS}) as well as the I^A, I^O and $Gm^{1,3,5,13,14}$ alleles. There are also a few low-frequency alleles (Gm^1, $Gm^{1,2}$, R^O and R^Z) which exist in higher proportions in one or two relatively isolated villages than elsewhere.

Previous hematological surveys on Bougainville largely corroborate these patterns, or do not contradict them. Kariks et al. (1957), in a large survey primarily of Nasioi- and Torau-speakers, found little variation among Nasioi in the ABO, MNSs, and Rh blood groups (the only systems tested). A shorter survey by Booth and Vines (1967) covers separate villages in three different language groups—Kunua, Buin, and Rotokas. The Buin results approximate the Siwai and Nasioi results of the survey by Kariks et al. and my own, but suggest that the northern Papuan-speakers may be even more diverse than even this survey of Eivo and two related villages indicates. The Harvard Solomon Island Project has sampled south Nasioi, Nagovisi, and Aita village concentrations fairly extensively, and these results will undoubtedly be of great interest when they are published. So far as I have seen them, they do not contradict the patterns outlined here.

Discussion

None of these studies affects the patterns established in the present series except in a corroborative way. The question remains: What is the proper explanation of these patterns? More specifically, what role has natural selection played in this differentiation?

I have assumed from the beginning that the intensity of a particular selection pressure over this small area is effectively uniform, and if it varied in intensity in the past, that it did so over the entire area in the same manner. Of course, this could easily be wrong.

The major stumbling block is that we still know very little about the nature of selection pressures which act on hematological polymorphisms, outside the malaria-related systems (Hb^s, Hb^c, the thalassemias, and G6PD deficiency). Some claim haptoglobin variants are also related to malaria. The ABO blood types are associated with different intestinal and stomach disorders, and some have looked for a link with smallpox resistance. But nothing is sure.

Using distributional arguments, others have suggested haptoglobin polymorphisms (Hp), immunoglobulin polymorphisms (Gm), and ABO have been subject to intense selection pressures in the past (e.g., Morton, Krieger and Mi, 1966; Workman et al. 1963). However, Cavalli-Sforza (1966) and Lewontin and Krakauer (1973) have argued that other systems, among them Kell, most HL-A alleles, and probably Gm appear to have been open to measurable selection.

In reviewing the frequency distributions from Bougainville from this vantage point, it is interesting that the ABO, Hp and Gm systems are the ones which vary markedly over the sample populations in a nonclinal fashion, as opposed to the Inv, Ss, and PHs frequency changes. In particular, the I^B, Gm^1, and

$Gm^{1,2}$ alleles are lost over the same range of villages, the Eivo-Simeku bottleneck. Hp^2 has significantly depressed values through this area, approaching the lowest frequencies reported from New Guinea or anywhere else (Kirk 1965). It is possible that the selection pressures needed to maintain low-frequency polymorphisms existed (or exist) throughout Bougainville, but in the sparsely populated Simeku-Eivo area the polymorphs were lost as the selection pressures were no longer adequate to counteract the tendency of such small marginal populations to become genetically uniform, or monomorphic. If this were the case, tendencies toward gene loss in this area should be manifest in most systems with low-frequency alleles. In the Bougainville data, the L^{NS} frequencies are very low, but have been described above as following a clinal pattern of variation. The L^{NS} results are not completely incompatible with this explanation, however, as fixation, or near fixation, occurs at the northern end of the bottleneck, so that the cline may be viewed as a dichotomy: in the north the Rotokas, Eivo, and Simeku together are very low in S, and in the south, the villages below the bottleneck have elevated S frequencies. The other two clinal distributions (Inv and PHs) involve high-frequency polymorphisms which are affected by the bottleneck only in having their steepest fluctuations there.

Population Structure Analysis

With the proliferation of known polymorphic serological systems and with the description of more and more populations, the need for a technique to compare populations with respect to many different polymorphisms has become all too plain. Not only would such a technique promise to facilitate population comparisons; it would reflect the possibility that the genetic loci investigated do not operate, and are not operated upon, singly.

The benefits of a valid multivariate treatment of large numbers of serological tests are clear enough, just as they are with multivariate analyses of anthropometric data. In this sense we have arrived at an analogous point in dealing with serological material with the biostatisticians of 1910–1930, who faced overwhelming amounts of biometric data for which univariate analyses were obviously inadequate and actually inappropriate.

There are possible objections to dealing with serological data in such fashion from both the biological and statistical viewpoints. All multivariate approaches offer at best nothing more than simplified, or more elegant, descriptions of differences between populations. For example, variation which is the product of natural selection cannot be separated from that resulting from any other cause. Variables will only be stressed by the degree to which they differ in frequency in the compared populations. This qualification holds for anthropometric and serologic analyses alike. Yet multivariate treatments of anthropometric data are now generally regarded as being appropriate and even mandatory (if not widely used).

Another biological argument that holds only for serological data is that because serological phenotypes are discrete, monogenetically controlled units of variation, they should be considered as such and remain uncombined. Certainly, they should be primarily discussed separately. But a proper treatment of the different patterns of allelic variation over multiple populations should, at the least, provide a more comprehensible picture of the nature of that variation.

Currently, there are two popular approaches to the description and analysis of genetic variation at multiple loci in human populations, both of which I will utilize. The first is referred to as "population structure analysis," following Morton (1968) and the second as "population distance." In his terminology, the population structure approach includes a consideration of the effects of all parameters on the mating patterns within a population. In other words, anything which modifies the expected uniform distribution of mates and genes within a population is a suitable subject for study in population structure. Information on population structure, then, comes from various sources, particularly from demography and from information on actual gene frequency distributions within a population. The quantitative statistics of population structure studies are coefficients of inbreeding and relationship, all of which are related to the variation, expected or observed, of the variance of the gene frequencies of the subpopulations around the total population mean.

As mentioned in Chapter 3, the usual measure of the variance when dealing with gene frequencies is called Wahlund's f, which has the formula,

$$f = \frac{\left[\sum_{i}^{N} (p_i - \bar{p})^2 \right] / N}{\bar{p}\,(1 - \bar{p})}$$

for a gene locus where there are only two allelic alternatives, p, and $(1 - \bar{p})$, and where there are N different subdivisions of equal size in the population. As a simple example, Tables 4.9a and 4.9b give a population divided into four equal groups, with gene frequencies for p of .30, .40, .60, and .70. The value in this case for f is .10 as opposed to 0.0 if all subdivisions had gene frequencies of .5. The other basic statistic in these cases is the covariance, which gives a measure of relationship of the different groups' patterns of variation. Its formula is

$$\text{cov}_{ij} = \frac{(p_i - \bar{p})(p_j - \bar{p})}{\bar{p}(1 - \bar{p})}$$

for any allele in subpopulations i and j. A negative covariance between two groups implies that their gene frequencies tend to vary in opposite directions from one another, while a positive one implies that, when one goes up, the other also goes up. An example is presented in the same table.

In simplest form, population structure analysis is primarily concerned with

Table 4.9a
Example of the Calculation of the Genetic Variance
$\sigma_p^2/\overline{pq}$ or Wahlund's f

	p_i	q_i	p_i^2	$2p_iq_i$	q_i^2
Group 1	.70	.30	.49	.42	.09
Group 2	.60	.40	.36	.48	.16
Group 3	.40	.60	.16	.48	.36
Group 4	.30	.70	.09	.42	.49
Total population	$\bar{p}$ = .50	$\bar{q}$ = .50	.275	.45	.275
Proportions if no subdivision			.25	.50	.25
Difference			+.025	−.05	+.025

$$\sigma_p^2 = \frac{\Sigma(p_i - \bar{p})^2}{N} = .10/4 = .025$$

$$f = \sigma_p^2/\overline{pq} = .025/.25 = .10$$

the variances of the different groups around the population mean, while the approach of population distance is primarily concerned with the relationships of particular subunits of the population, and hence with the covariances as well. Gene frequencies can be converted to genetic distances in a variety of different fashions, with those populations which are most similar in their gene frequencies (or having the highest covariance) having the smallest "distance" between them. The same measure described in Chapter 3, $d_{ij}^2 = \text{var}_{ii} + \text{var}_{ij} - 2\,\text{cov}_{ij}$, will be used here. This also happens to be equal to

$$\sum_i \frac{(p_i - p_j)^2}{\bar{p}},$$

which is a common measure of "genetic distance," G, proposed by Balakrishnan and Sanghvi in 1968. I will utilize both approaches as they complement one another quite well. As shown in Appendix I, genetic distance statistics can be written as functions of variances and covariances, the stuff of population structure, and as shown in Chapter 3, the variances and covariances may be predicted by demographic (migrational) information, although with a slightly altered interpretation.

Harpending and Jenkins (1973) have introduced an "unbiased Wahlund's f" which takes account of bias introduced by unequal sampling procedures. When all individuals in a village are included, then the gene frequencies should

Table 4.9b
Calculation of Genetic Covariances

$$\text{Covariance}_{ij} = [(p_i - \bar{p})\,(p_j - \bar{p})]\,/\bar{p}\bar{q}$$

$$\text{cov}_{12} = (.2)\,(.1)/(.5)\,(.5) = .08$$

$$\text{cov}_{13} = (.2)\,(-.1)/(.5)\,(.5) = -.08$$

$$\text{cov}_{14} = (.2)\,(-.2)/(.5)\,(.5) = -.16$$

$$\text{cov}_{23} = (.1)\,(-.1)/(.5)\,(.5) = -.04$$

$$\text{cov}_{24} = (.1)\,(-.2)/(.5)\,(.5) = -.08$$

$$\text{cov}_{34} = (-.1)\,(-.2)/(.5)\,(.5) = .08$$

be weighted solely by the size of the sample. But, as is inevitably the case, when the sample is only a portion of the census size, then there is some error in the estimation of the true gene frequency of the village, so this sampling bias must be taken into account as well. Hence, Harpending's formula for the unbiased village variance is

$$r_{ii} = \frac{i}{\text{alleles}} \sum_{k} \left[\frac{(p_{ik} - p_k)^2}{p_k\,(1 - p_k)} - \frac{1 - \dfrac{\text{sample}_i}{\text{census}_i}}{\text{sample}_i} \right]$$

and

$$r_{ii} = \sum_{i} w_i r_{ii} \quad ,$$

where i = the ith population,
  k = the kth allele,
  p = gene frequency, and
  $w_i = n_i/N$.

Aside from the differences in weighting, the results should be identical.

The estimates for r_{ii}, or the variance of the village gene frequencies around the total sample average, or Wahlund's f, are presented in Table 4.10. There are, obviously, different estimates for different genes and different genetic systems. As expected from the earlier consideration of the individual results, the

Table 4.10
Estimates of Wahlund's $\bar{f}_{st}$, or $\bar{r}_{ii}$,
over Seven Blood Polymorphisms

ABO	.0522
Rh	.0113
Gm	.0767
Inv	.0777
Hp	.0563
PHs	.0490
MNSs	.0430
Average	.0477

Inv alleles have the largest variation over this area, followed by the Gm, Hp, ABO, MNSs, PHs, and finally, the almost uniform Rh gene frequencies. Although the highest estimated variance, i.e., for Inv, is more than eight times larger than the lowest, for Rh, statistically the differences in the variances for the different systems are not significant (see Friedlaender 1971a).

This homogeneity of estimates from the different polymorphisms suggests that rates of migration and isolation can account for the pattern of variation within the total sample, because migration as such should affect all genetic loci equally. That is, if the agent responsible for maintaining homogeneity of gene frequencies is long-range migration, then all alleles should provide estimates of the **R** matrix with identical expectations.

There is a great deal of variety, however, in the people who have come into the population to marry, which violates the assumption that the in-migrants come from a homogeneous population with a particular gene frequency. Such a pattern of admixture should lead, and in this case does lead, to higher Wahlund f values than those predicted by the uniform systematic pressure of the migration matrix theory used in Chapter 3, since the immigrants in different areas come from parent populations with very different gene frequencies.

From the results in Table 4.10, we might infer either that donor populations in the north have high frequencies of Inv^1, for example, or that systematic selection pressure is at a minimum on this system, while the opposite might hold for the Rh system.

In his later papers (1948, 1951), where he extended the concept of genetic drift to all chance fluctuations that could influence gene frequencies, Wright showed that whenever the systematic pressures (migration, selection, and mutation) were smaller than $1/2N^e$ (where N^e is the effective population size), then the effects of drift or sampling effects could be significant. For example, in the case of village 12, Pomaua, with an effective population size of approximately 50, sampling events, or the effects of drift, should be significant and important in the face of contrary selection pressures with coefficients less than .01.

It is clear from the preceding discussion that pooling the results from different genetic systems essentially ignores the possible reality of selection pressures acting at different intensities on different genetic systems. However, I intend to do just that in an effort to search for overall patterns of variation which may be tied to the migrational dynamics within the population. The entire attempt to find a "genetic distance" between pairs of groups involves such simplifications, and this should be acknowledged from the outset.

As in Chapter 3, I have chosen to represent the "genetic distance" relationships of the villages by subjecting the **R** matrix of blood polymorphism gene frequency variances and covariances to a principal components analysis. Through this procedure, similar patterns of variation in different alleles can be identified and grouped together, forming different "axes" of variation, which, taken altogether, will give the "average" genetic distance between the groups. In the present instance, this involves searching the matrix of relationship coefficients (or group variances and covariances) for those genes which vary the most, and in the same fashion, over the 18 villages. After their variation is taken account of in a first component, the remaining variation and covariation is again searched for the most important variables, and so forth, until no significant variation is left. A worked example of a principal components analysis of an **R** matrix is given in Harpending and Jenkins (1973). This technique, used in a similar fashion by Workman et al. (1973) and Morton and Lalouel (1972), was applied to these data. The results are illustrated here in Tables 4.11 and 4.12 and Figure 4.1.

The 18 groups are distributed in a pattern which reflects their geographic and linguistic relationships, evidently more clearly than the results of the migration data manipulation. The first vector, or principal component, separates the total sample into two groups, the northern and the southern. All the villages north of the Nasioi (villages 1–10) except for Boira (9) have a positive coefficient on this component, while all south of the Nasioi-Simeku border have a negative score. Above all, villages 1 and 2, the Rotokas and Aita, are placed at an extreme position on this most important component (accounting for .353 of the total variance), while most other village samples show fewer marked differences from one another.

The coefficients of the different genes associated with this first component show that this distribution is associated with or caused by those alleles which, from the preceding discussion, we would expect. Both Inv genes, L^{NS}, I^B, and P^a and P^b are strongly associated with this axis. These are the "clines" in the blood polymorphs.

The second component, which accounts for substantially less variation than the first, reflects the "U-shaped" variation discussed earlier. That is, this component groups samples 1 and 2 with the Melanesian-speaking Rorovanas and Arawas (villages 17 and 18). This arrangement, not surprisingly is characteristic of the haptoglobin genes as well as I^B.

The third component, which is only slightly less important than the second, contrasts the first sample from Aita with all the rest—in particular the Rotokas (sample 2). This is a component strongly influenced by the variation in Gm alleles. The other components are much less important and hard to interpret.

By plotting the first three axes against one another (Figure 4.1), we can get a clearer picture of the overall clusterings of the villages on these summary components. The 18 groups are distributed in a pattern which reflects their geographic and linguistic relationships in a fairly clear fashion, certainly more

Table 4.11
Principal Components Analysis of the
Blood Polymorphism Relationship Matrix

Villages	Component scores		
	1	2	3
1	.35	−.22	.62
2	.58	−.34	−.36
3	.06	.18	.10
4	.15	.34	−.01
5	.18	.25	.21
6	.06	.27	.29
7	.16	.17	−.18
8	.02	.10	−.23
9	−.15	.36	−.01
10	.09	.21	−.35
11	−.27	.22	−.08
12	−.07	−.19	−.24
13	−.25	−.02	.08
14	−.31	−.14	.02
15	−.25	.00	.03
16	−.07	−.11	−.17
17	−.28	−.30	−.09
18	−.19	−.36	.20

clearly than the results of the migration data manipulation. They also suggest that the variation in the less populous northern sector of the sample area is more extreme than in the south, even over considerably greater geographic distances.

The most remarkable aspect of the analysis is that by pooling results from seven different blood polymorphisms, a pattern emerges which is anything but surprising in view of what is known about the past and present relationships of the villages. Just how close a correspondence there is will be considered in more detail in the concluding chapter. The principal vector tends to make distinctions among the northern groups as opposed to the more homogeneous southern groups, with villages 1 and 2 especially remote. Generally speaking, the distinctions are greater in the blood analysis than in the migration results

Table 4.12
Relative Importance of the Blood Polymorphism
Alleles on the First Three Components

	Component weights		
Alleles	1	2	3
Gm^1	.81	−.73	−1.21
$Gm^{1,2}$	−.12	−.90	1.20
$Gm^{1,3,5,13,14}$	−.62	.81	.67
R^1	−.49	.31	−.24
R^2	.50	−.28	.23
R^0	−.14	−1.03	.39
Hp^1	.06	.77	−.20
Hp^2	−.06	−.77	.20
I^A	.52	−.07	.40
I^O	−.28	.41	−.51
I^B	−1.02	−1.72	.72
L^{NS}	.30	.32	.03
L^{NS}	−1.43	−.68	−.32
L^{MS}	.18	−.16	.07
L^{MS}	−.41	.98	1.05
Inv^{1+}	.81	.12	.37
Inv^{1-}	−.81	−.12	−.37
p^a	.72	.63	.20
p^b	−.72	−.63	−.20

and seem to follow language and geographic relationships more closely. In particular, the Melanesian-speaking beach people, the Torau and Arawas (17 and 18), are more clearly separated in the blood analysis than in the migration results.

Heterozygosity

A related approach to the description of blood genetic variation might be called the calculation of absolute heterozygosity and derives from the work of Lewontin (1967, 1972) and Harris (1969). The Wahlund approach used in the previous section treats relative heterozygosity, or the decrease in heterozygosity due to population subdivision over what is expected in a random mating situation. Lewontin and Harris are interested in what proportion of the gene

4.1 Blood polymorphism relationships of the 18 samples: principal components analysis of the blood polymorphism relationship matrix

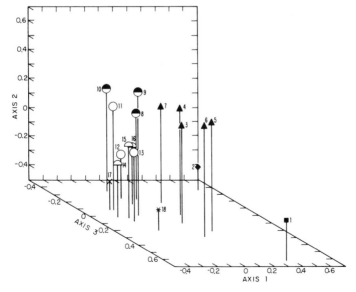

loci analyzed yields actual evidence of polymorphisms, and at how many loci in a given individual one might expect to find heterozygous states.

For instance, the heterozygosity of a population (d) at a particular locus should be

$$d = 1 - \sum_{k=1}^{t} p_k{}^2,$$

where t is the number of alleles at the locus and p_k is the frequency of the kth allele. Then the heterozygosity (D) of a population over n different loci is computed as the average over all loci:

$$D = \frac{1}{n} \sum_{i=1}^{n} d_i.$$

Nei and Roychoudhury (1973) have derived bias and sampling variance for heterozygosity. It is on the order of $1/2n$ and so is relatively small for samples over 20.[5]

Table 4.13 presents the values of d for each village in the sample, the average village values (d_{village}), the values for each language group which is represented by more than one village (D_{language}), and the value for the average gene frequencies over the total sample $(D_{\text{Bougainville}})$. If the fixed genes of the Duffy, Kell, and transferrin systems were added, the absolute values for heterozygosity would be lowered considerably, but the relative heterozygosities of the different groups would remain unchanged. Although there is a wide variance in d estimates for the village samples (from .269 to .390), it is difficult to find much that is revealing in the distribution. Fluctuations in the I^B gene and in $Gm^{1,2}$ tend to make the extreme northern and southern samples appear more heterozygous, and the central ones less so, but statements beyond this are unwarranted.

Comparisons with other reports are also hazardous (e.g., Harpending and Chasco 1975, Lewontin 1972), as each depends upon a different battery of blood polymorphism tests and comes from a different area of the globe, each sometimes characterized by widely divergent sets of alleles. It appears that from

5. Actually, Lewontin prefers to use a closely related statistic, Shannon-Weaver information (see Lewontin 1972), but d suffices here.

Table 4.13
Diversity by Group

Village	$D_{village}$	Language Group	$D_{language}$
1. Nupatoro	.3825	Eivo	.3399
2. Okowopaia	.3801	Simeku	.3304
3. Kopani	.3256	Nasioi	.3250
4. Kopikiri	.3255	Siwai	.3227
5. Nasiwoiwa	.3284		
6. Atamo	.3368	$\bar{D}_{language} = .3295$	
7. Uruto	.3558		
8. Karnavitu	.3408		
9. Boira	.2982	$\bar{D}_{Bougainville} = .3391$	
10. Korpei	.3395		
11. Sieronji	.2692		
12. Pomaua	.3470		
13. Bairima	.3165		
14. Turungum	.3143		
15. Moronei	.3179		
16. Old Siwai	.3542		
17. Arawa	.3581		
18. Rorovana	.3910		
$\bar{D}_{village}$ =	.3379		
$\bar{D}_{14 \, villages}$ =	.3264		

a system-by-system comparison of d values the heterozygosity of the Bougainville samples is relatively low, as one would expect from the isolated nature of the samples and the presumably small size of the original founder populations. Table 4.14 shows that, as expected, it is the Inv, ABO, and Hp systems which account for most of the heterozygosity within the sample, while Rh is lowest of the segregating systems. There are many possible implications which others may choose to draw from this table concerning the possible effects of differences in selection pressures, but that seems premature.

Following Lewontin's general approach, the heterozygosity within the sample can be partitioned in much the same way as with the Wahlund technique with average village, language group, and village-within-language group heterozygosi-

Table 4.14
Proportion of Genetic Diversity Accounted for
within and between Villages and Language Groups

Gene systems	Total $D_{\text{Bougainville}}$	Village	Among village within lang.	Among lang. group
ABO	.4191	1.0169	-.0066	-.0103
Rh	.1529	.8571	.0771	.0658
MNSs	.3696	1.0478	-.0299	-.0178
PHs	.3393	.7966	.0186	.1848
Hp	.4015	.9210	.0896	-.0105
Gm	.1987	.9665	.0264	.0071
Inv	.4924	.9654	.0184	.0162
Average	.3391	.9626	.0091	.0283

ties presented as proportions of the maximum heterozygosity, which the average gene frequencies over the total sample should yield.[6]

Although there is a great deal of variation from one system to the next, the results suggest that the heterozygosity within villages is disproportionately large and that the reduction in absolute heterozygosity due to population subdivision is relatively small (an average of less than 4 percent in segregating loci). Also, there is considerable isolation by distance or by language group. Diversity among villages within language areas is approximately one third that among language groups.

All these different approaches suggest essentially the same thing: while there is considerable genetic variability within local village areas, what variation there is from one village to the next is principally organized around language and geographic relationships. At least for these groups, taxonomies based on language similarities would seem as valid as any.

6. The proportions are calculated as follows: using only those language groups where more than two villages were sampled (14 villages),

Within-village proportion $= D_{14\text{ villages}}/D_{\text{Bougainville}}$

Among village-within-language group $= (D_{\text{Language}} - D_{14\text{ villages}})/D_{\text{Bougainville}}$

Among Language Groups $= (D_{\text{Bougainville}} - D_{\text{Languages}})/D_{\text{Bougainville}}$

5 Anthropometry

If I was sure of one characteristic of the Bougainville population from the out-set, it was that anthropometric variation existed in abundance. Douglas Oliver, ever the conscientious ethnographer, managed to measure over 1,300 males in late adolescence or adulthood in 1938–39 as a diversion in his field routine. He covered a large proportion of the male population in the Siwai and Nagovisi regions, but other language groups of the island were represented primarily by plantation workers he encountered on walks around the coast. From the pub-lished reports (Oliver 1955; Oliver and Howells 1957, 1960; Howells 1966a) it was clear that variation within language groups was substantially less than among the various language areas. He concluded that there were three extreme "types" represented on the island—robust, tall northern mountaineers; tall but thin beach people; and short, bracycephalic southern mountaineers. However, attempts to establish physical relationships among the language groups he sur-veyed were less than completely satisfactory because of the sometimes very small and select representative samples.

Another contention of Oliver's was that in 1939 there was some suggestion that the younger adult males were growing taller and larger than their fathers, or that something similar to the well-known secular trend in Western Europe and North America was evident in this population, presumably caused by some effect of early Western influence.

Both of these points bear on the present survey results. First, I had to establish the anthropometric differences within the section of the island popula-tion I was studying intensively, trying to maintain a more complete sampling procedure than Oliver had been able to follow over larger portions of the island. Secondly, I had to establish the nature of the anthropometric variation over the adult age cohorts, for two reasons. First, population heterogeneity, presumably due to genetic causes, might be confounded with aging or secular trend differ-ences, and to account for these was important. Second, the effect of the im-portant differences in the health and nutritional environment in the last thirty years might be reflected in the cross-sectional results, and this in itself was interesting. Here, too, was the opportunity to follow up and remeasure some of the surviving men measured almost thirty years before by Oliver, an unparal-leled opportunity to establish the existence of a secular trend in a traditional society, implied in his cross-sectional results.

Data Collection

Anthropometrists have invented literally hundreds of different measurements to be made on the human body, but time and again, particularly in multivariate studies, it appears that, in describing physical variation, a few measurements provide very clear distinctions among different groups. As my primary goal was a multivariate analysis of village and language group variation within the sample, I was strongly influenced in the choice of measurements by previous multivariate studies on living human populations. The eight measurements used in Penrose's "Size and Shape" computations (1954) formed the core of the battery. These were also used because Howells, in dealing with Oliver's more extensive test battery on Bougainville males, chose these measures to provide for

24. Class Reunion: Some of the men of northeast Siwai measured in 1938 and again in 1967

25. Kakai exhibiting photographs Oliver took of him as a young man in 1938

his calculation of Mahalanobis' D^2 statistic without correction for intercorrelations. That is, from his own analysis of Oliver's measurements, and from his studies on the correlation of various measurements on the living, Howells (1951) decided upon these variables as the ones which would give an adequate picture of physical differentiation, and which at the same time were not strongly intercorrelated. These eight measurements were: sitting height, upper extremity length, chest breadth, head length, head breadth, minimum frontal breadth, bigonial breadth, and total face height. Weight and stature (height) were added to this battery, the rationale being that these standard measures are so commonly taken that they should be included in any study, and that they might reflect nutritional or health differences between groups. In retrospect, skin fold

26. Tumare holding his photographs as a 20-year old

measures would have been a good addition to the battery, but weight should reflect skin fold variation. The additional facial measures of nose breadth, nose height, and bizygomatic breadth were included because the face seems to be the area most sensitive to population differentiation, and the taking of these measurements consumed little extra time.

Measuring was restricted to adult males for a number of reasons.[1] The great bulk of comparative data available, and particularly Oliver's long series, was taken from male subjects. Since part of the envisioned study involved a re-measuring of as many as possible of the people he had measured twenty-eight years before, males had to be included. A female series, besides adding a great time burden, would probably have added little to an estimation of physical differences between groups. Only "adult" males, arbitrarily defined as those men with fully erupted third molars, were measured. Younger individuals could have served for comparison of cross-sectional growth curves, but sufficient numbers in each age and linguistic category for such a study would have been hard to obtain.

With these stipulations I tried to measure all adult males who were regarded by the villagers as present residents of the villages within the survey (see Plate 21 and 23).[2] The sample sizes of each village, along with the linguistic groupings, are given in Table 5.1. The average size of the village samples is about 30, but two groups, from Kopikiri and Sieronji, are much smaller than the rest, and this affects the statistical results, both univariate and multivariate. Group 16 is different in character from the rest, being composed of old men who had been measured by Douglas Oliver in 1938–39 from northeast Siwai villages other than Turungum (14) and Moronei (15). They are included here as a test for the effects of age on population comparisons, as all the other samples include adult males of all ages.

1. Age estimates among the Siwai are Oliver's own. Undoubtedly this makes for errors, but by using five-year categories as a minimum unit in the analysis, I doubt I am making serious errors. In the Rorovana and Arawa villages, corroborating ages for men were available from the mission baptismal records.

2. There were accordingly a few off-islanders who were measured, but special note of them was made so that I could decide at a later time whether or not they should be included. Because over 50 percent of them were singled out as deviants by the Churchill Editing Program discussed below, I eventually decided not to include them in the anthropometric calculations.

Aging and Secular Trend Effects

It was principally because of Oliver's earlier study that I included northeast Siwai in my 1966–67 survey, for this provided me with the opportunity to carry out a longitudinal study of aging effects in Bougainville males. To my knowledge no such study has been carried out on any other nonurbanized population. I surveyed two villages where Oliver had lived for extended periods of time (Turungum and Moronei, numbers 14 and 15) and also sent word to other villages of the northeast that I wanted to measure and take blood from anyone who had been measured by Oliver twenty-eight years earlier. In this way I was able to include 45 Siwai in a follow-up survey—17 from the two

Table 5.1
Anthropometric Samples and Sample Sizes

	Linguistic affiliation	Village	Adult males measured
Northern NAN	Aita (Rotokas dialect)	1. Nupatoro	21
	Rotokas	2. Okowopaia	19
	Eivo	3. Kopani	54
	Eivo	4. Kopikiri	10
	Eivo	5. Nasiwoiwa	36
	Eivo	6. Atamo	32
	Eivo and Simeku	7. Uruto	30
Southern NAN	Simeku (Nasioi sub-language)	8. Karnavito	35
	Simeku	9. Boira	25
	Simeku	10. Korpei	50
	Nasioi	11. Sieronji	9
	Nasioi	12. Pomaua	34
	Nasioi	13. Bairima	21
	Siwai	14. Turungum	19
	Siwai	15. Moronei	32
	Siwai	16. Old Siwai[a]	28
AN	Uruava (previously)	17. Arawa	20
	Torau	18. Rorovana	54

[a]These are men from northeast Siwai villages who were measured thirty years previously by D.L. Oliver and were remeasured by the author in 1967.

villages covered intensively and 28 from the immediately surrounding area. Their ages varied from forty to over seventy in 1967. This last group is the "Old Siwai" sample, number 16 in the series. Identification of remeasured individuals was not an insuperable task. Oliver had taken photographs of all the men he measured, and these were available to me in the field, along with the original measurement blanks (see Plates 27–62). Only four men who claimed to have been measured by Oliver were finally excluded from the tabulations, either because they could not be positively identified with photographs or because their old measurement records could not be found in Peabody Museum upon my return. Oliver had also visited Arawa and Rorovana villages (numbers 17 and 18), which I also surveyed, so that, in a similar manner, I located and measured 17 additional men from that very different area of the island. Undoubtedly, if I had wished, I could have located many more men previously measured by Oliver, especially in Siwai and Nagovisi. About half of Oliver's 1,300 subjects lived in these two language areas, which were sampled fairly exhaustively.

Table 5.2
Height—Cross-Sectional Results of
Oliver's (1939) and Friedlaender's (1967) Surveys

Age	Oliver			Friedlaender			$\overline{\Delta}^b$
	Mean height (Cm)	σ^a	N	Mean height (Cm)	σ^a	N	
10–14	148	7	4				
15–19	158	7	16				
20–29	162	6	95	164	4	7	−2
30–39	161	5	83	164	6	19	−3
40–49	161	6	56	162	6	15	−1
50–59	161	6	29	162	7	31	−1
60–69	160	3	5	157	4	5	4
70–80				155	7	3	
Total			288			79	

$^a\sigma$ = standard deviation.
$^b\Delta$ = the mean differences.

Table 5.2 and Figure 5.1 compare the cross-sectional results of Oliver's and my surveys for stature in northeast Siwai, the area most intensively covered by Oliver, and with the most overlap in both surveys. Oliver's series is remarkably uniform over the adult cohorts, the 20–29-year cohort being a centimeter taller than the succeeding three. Quite to the contrary, my smaller series shows

5.1   Stature: cross-sectional comparisons

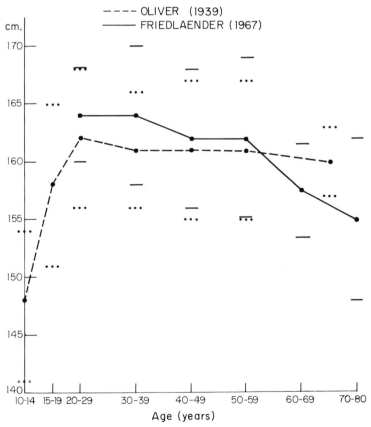

a pronounced difference in stature in those adult males younger than 49 as opposed to those 50 and over, who average at least 2 centimeters shorter. It should be noted that the oldest two cohorts have such smaller numbers that their interpretation is questionable in a cross-sectional comparison. Note that there is no difference between the average stature of the 20–29-year cohort in Oliver's series and my 40–59-year cohorts, which include many of the same individuals.

The most obvious preliminary interpretation of this comparison is that there has been a secular trend in Siwai populations, if not elsewhere on Bougainville, and that this increase may well have begun initially among those males born during the years 1910–1919, and subsequently increased another 2 centimeters. It is also possible that Oliver and I measured our subjects in different ways or had a significantly different set of subjects in ways I did not realize. Differential mortality with regard to height, with taller people dying earlier, might be one possible explanation.

Figure 5.2 and Table 5.3, with the restudy results of the 62 subjects measured both in 1938–39 and 1967, largely answer these questions. These men, 45 from Siwai and 17 from Rorovana and Arawa, have been grouped according to their ages in 1938–39. The dotted line in the figure represents the average values of this select group in 1938–39, and the solid line gives their height 28 or 29 years later, in early 1967. The empty circles are the cross-sectional values for the corresponding classes in Oliver's study taken from the previous figure for reference.

In Figure 5.2 the two lines cross, indicating that the men Oliver measured who were younger than 25 were generally taller in the restudy, while those 25 and older when Oliver measured them (and who were 53 and older when remeasured) generally had lower values the second time around. These differences grow from .7 to 2.8 centimeters with each age class. I believe this graph indicates that there is little difference in the measuring technique used by Oliver and me. The results from men who were fully grown in 1939 and who were still in their early fifties in 1967 correspond quite closely, indicating that the secular trend explanation for the differences in the cross-sectional graph is the correct one. Moreover, there is good evidence for real "shrinkage" beginning in the late fifties and increasing thereafter.

Furthermore, if we compare the dotted line with the open circles, it is clear that the 62 men were fairly close to the average values for their age classes in 1939, or, if anything, slightly taller. Therefore, differential survival of shorter men into later life does not seem to be a likely explanation for the sharp drop in stature for later age classes.

5.2   Stature: longitudinal changes. (O indicates the mean for the total cohort measured by Oliver in 1939; also given in Figure 5.1)

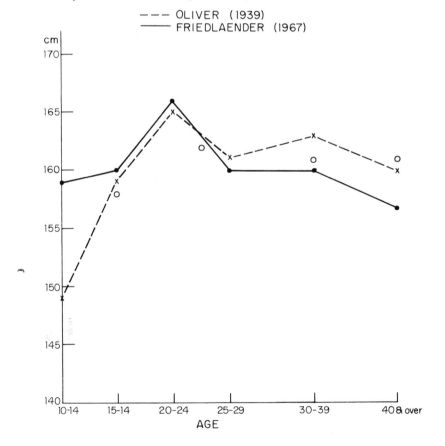

Table 5.3
Heights of Individuals Measured 28 Years
Apart by Oliver (1939) and Friedlaender (1967)

Age in 1938	Oliver (1939)		Friedlaender (1967)			
	Mean height (Cm)	σ	Mean height (Cm)	σ	N	Δ̄
10–14	147.7	6	159.2	2	3	–11.5
15–19	157.5	8	160.3	7	6	–2.8
20–24	164.8	6	165.8	6	15	–1.0
25–29	161.1	7	160.4	7	22	.7
30–39	163.1	5	160.5	5	12	2.6
40 and over	159.5	8	156.7	7	4	2.8
Total					62	

Table 5.4
Upper-Extremity Length—Cross-sectional Results
of Oliver's (1939) and Friedlaender's (1967) Surveys

Age	Oliver (1939)			Friedlaender (1967)			
	Mean arm lgth. (Cm)	σ	N	Mean arm lgth. (Cm)	σ	N	Δ̄
10–14	67.3	2.7	4				
15–19	73.5	4.5	16				
20–29	75.6	3.3	94	77.8	3.3	7	–2.2
30–39	75.0	4.5	83	76.6	3.9	19	–1.6
40–49	74.7	4.0	56	75.6	2.9	15	–.9
50–59	75.6	4.2	29	76.2	4.2	30	–.6
60–69	75.4	2.3	5	74.1	2.5	5	2.7
70–80				70.3	1.4	3	
Total			287			79	

The other measured trunk and limb lengths (upper extremity length and sitting height) given in Tables 5.4–5.7 and Figures 5.3–5.6 suggest that by far the largest portion of the "shrinkage" occurs within the trunk, but that the younger generation attains longer trunk and limb lengths. Measurement error is a distinct possibility here, however.

Table 5.5
Upper-Extremity Lengths of Individuals Measured
28 Years Apart by Oliver and Friedlaender

Age in 1938	Oliver (1939)		Friedlaender (1967)			
	Mean arm lgth. (Cm)	σ	Mean arm lgth. (Cm)	σ	N	$\overline{\Delta}$
10–14	68.6	1.3	73.5	.9	3	−4.8
15–19	72.2	4.2	75.1	3.4	6	−2.9
20–24	76.7	3.8	77.7	4.1	15	−1.0
25–29	74.8	3.5	75.9	4.1	21	−.4
30–39	75.8	3.5	76.0	3.1	12	−.7
40 and over	75.2	5.9	75.5	4.8	4	−.3
Total					61	

Table 5.6
Sitting Height—Cross-sectional Results of
Oliver's (1939) and Friedlaender's (1967) Surveys

Age	Oliver (1939)			Friedlaender (1967)			
	Mean sitting ht. (Cm)	σ	N	Mean sitting ht. (Cm)	σ	N	$\overline{\Delta}$
10–14	75.1	3.2	4				
15–19	80.5	4.9	16				
20–29	82.5	3.7	95	84.7	1.1	7	−2.2
30–39	82.5	3.4	83	84.7	2.8	19	−2.2
40–49	82.7	4.3	55	82.3	3.0	15	.4
50–59	81.6	4.7	29	83.0	4.1	31	−1.4
60–69	77.1	8.5	5	80.2	1.9	5	−1.8
70–80				76.8	3.4	3	
Total			287			80	

Table 5.7
Sitting Heights of Individuals Measured
28 Years Apart by Oliver and Friedlaender

	Oliver (1939)		Friedlaender (1967)			
Age in 1938	Mean sitting ht. (Cm)	σ	Mean sitting ht. (Cm)	σ	N	Δ̄
10–14	74.3	4.8	82.4	2.6	3	−8.1
15–19	80.4	5.5	81.5	4.2	6	−1.1
20–24	83.1	6.1	85.6	2.9	15	−2.5
25–29	82.1	3.5	82.8	3.5	22	−.7
30–39	83.8	2.8	81.8	2.3	12	2.0
40 and over	81.8	3.2	78.4	4.4	4	3.4
Total					62	

5.3 Upper extremity length: cross-sectional comparisons

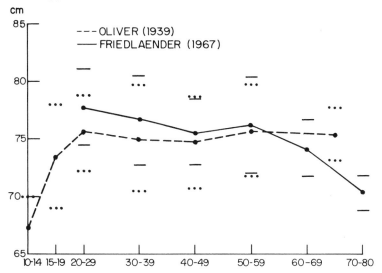

5.4 Upper extremity length: longitudinal changes

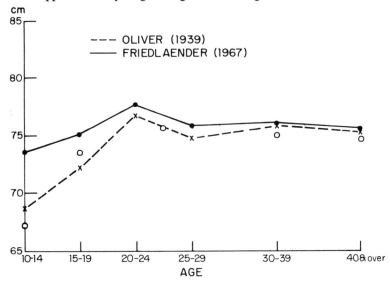

5.5 Sitting height: cross-sectional comparisons

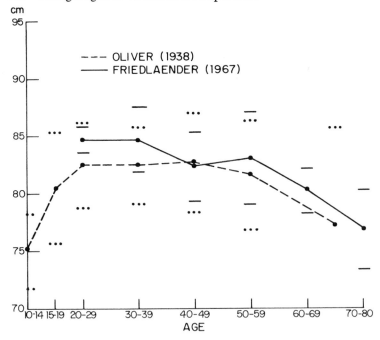

The cross-sectional curves for weight (Figure 5.7, Table 5.8), a variable which reflects many different crosscurrents and influences, are generally similar to those for stature, although they are somewhat more pronounced. Except for those men in their fourth decade, the values for the present survey cohorts are all about 5 pounds greater than for the corresponding cohorts in Oliver's survey.

The longitudinal results, presented in Figure 5.8 and Table 5.9, rule out the possibility that these differences are attributable to our using different scales. The reweighed men, except for the youngest category, tend to be lower in weight than they were in 1938–39, as opposed to the cross-sectional study, where the newer means for most groups are higher.

As with stature, the most reasonable interpretation is that there has been an increase in the maximum attained weight in the new postwar generation, and that there has been a decrease in weight in later life, although exactly how soon this decrease begins is not entirely clear.

5.6   Sitting height: longitudinal changes

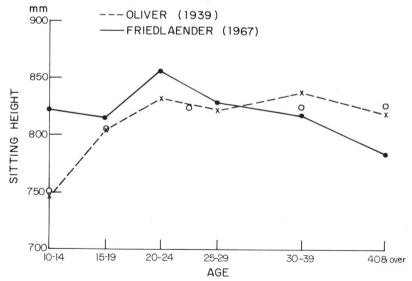

5.7   Weight: cross-sectional comparisons

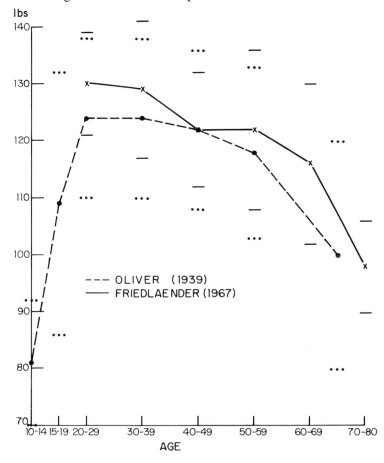

Table 5.8
Weight—Cross-sectional Results of
Oliver's (1939) and Friedlaender's (1967) Surveys

Age	Oliver			Friedlaender			
	Mean weight (lbs.)	σ	N	Mean weight (lbs.)	σ	N	$\bar{\Delta}$
10–14	81	11	4				
15–19	109	23	16				
20–29	124	14	94	130	9	7	–6
30–39	124	14	83	129	12	19	–5
40–49	122	14	56	122	10	15	0
50–59	118	15	29	122	14	30	–4
60–69	100	20	5	116	14	5	–9
70–80				98	8	3	
Total			287			79	

There is a technical question here, as two different scales were used, but again in this instance, the longitudinal results make large differences unlikely. Figure 5.8 and Table 5.9 show that the reweighed men tend to be fairly consistently lower in weight than they were in 1939, as opposed to the cross-sectional results, where the newer means for all groups are higher. In the restudied sub-sample, the average difference between the first and second weighings increases gradually up to 10 pounds for men now 68 to 72 years old, a large weight loss by any reckoning.

As with stature, the most reasonable interpretation is that there has been an increase in the maximum attained weight in the new postwar generation, and that there has been a decrease in weight in later life, which appears to take place earlier than does the shrinkage in stature. Vines (1970), interpreting cross-sectional studies in New Guinea, believes this weight loss to be most pronounced and earliest in females even in their mid-thirties.

This pattern is, of course, contrary to the pattern for industrialized populations, where the later years of adulthood are generally characterized by weight gain, although variation from individual to individual in this respect is great.

The results from other measurements taken by Oliver, and included in the battery are less interesting. There is good evidence for the so-called "senile chest"; that is, chest breadth decreasing sharply with increasing age, and chest depth, at least in this case, increasing only slightly.

In the head there are only some minor differences between curves in both cross-sectional and follow-up studies. Nose length and breadth tend upward in later life, but total face height declines, most likely because of tooth loss.

5.8   Weight: longitudinal changes

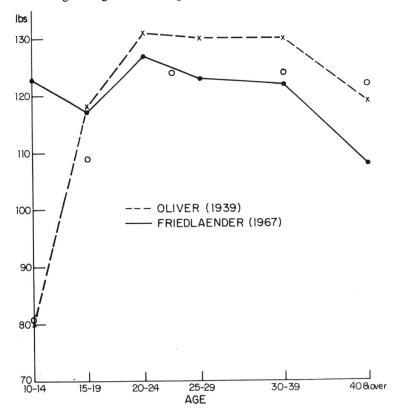

In summary, then, there is good evidence for "shrinkage," particularly in the vertebral column, and for an accompanying loss in weight. This shrinkage occurs somewhat earlier in life than seems the case in modern industrialized societies. It is also clear that at least the postwar generation of Siwai and Torau and Arawa males are bigger than their fathers by an appreciable amount—by at least a centimeter. The cause of this increase is most likely improved nutrition. If these age-associated differences have any effect in the population comparisons, they would be toward making the villagers living closest to Kieta (villages 17, 18, 12, and 13) largest and heaviest, with the northerners the smallest.

There is the distinct possibility that the differences in the cross-sectional curves may be at least partially due to differential survival of men in the 28-year interval between 1938–39 and 1967. Note that, particularly in weight (Figure 5.8), the 61 "surviving" males were, on the average, more than 5 pounds heavier than the mean for their entire cohort in 1938–39. To explore this matter I have compared "recovered" and "unrecovered" groups of both Siwai- and Melanesian-speakers by ten-year age cohorts.

In northeast Siwai, where the sampling was very close to complete in both 1938 and 1967, almost half of the teenagers, and almost a third of the men in their twenties who were measured by Oliver, were recovered and measured in

Table 5.9
Weights of Individuals Measured
28 Years Apart by Oliver and Friedlaender

Age in 1938	Oliver		Friedlaender			
	Mean weight (lbs.)	σ	Mean weight (lbs.)	σ	N	$\bar{\Delta}$
10–14	80	13	123	2	3	−43
15–19	118	22	117	17	6	1
20–24	131	16	127	15	15	5
25–29	130	21	123	17	21	8
30–39	130	18	122	15	12	8
40 and over	119	15	108	17	4	11
Total					61	

1967 (see Table 5.10). For the smaller series of Melanesian-speakers from Rorovana and Arawa, the recovery rate was better for men in their twenties and much better for men in their thirties. Almost no men in their forties in 1938–39 were recovered in 1967.

Comparing the "recovered" and "unrecovered" males for the two language areas, I have found, among other things, that the "recovered" males in both Siwai- and Melanesian-speaking groups do tend to outweigh the "unrecovered" groups in the early decades. This is reflected in Figure 5.8, although only Siwai teenagers show this clearly, while Melanesians up through their thirties have this pattern. In the fourth decade (and also third for Siwai), it was the lighter male who tended to "survive" to be remeasured at age 70 or so. As for differences in height, these are less distinctive, but they parallel the pattern of weight differences in both groups. All of these results are more fully covered in Fried-laender and Oliver (1975).

All this suggests that the taller, and especially the heavier young men, survived the war and the rest of the intervening 28 years between 1938–39 and 1967 in greatest numbers, but that the very few middle-aged men who attained age 70 or thereabouts in 1967 were shorter and lighter than their age mates in

Table 5.10
Recovery Rate of Northeast Siwai
and Rorovana and Arawa Males

Age	Siwai		Melanesian-speakers	
	Measured by Oliver	Recovered in 1967	Measured by Oliver	Recovered in 1967
10–14	5	2	0	0
15–19	15	6	6	0
20–24	43	11	14	4
25–29	52	17	10	5
30–34	49	5	9	2
35–39	32	1	11	4
40–44	31	2	8	2
45–49	24	0	5	0
Total	251	44	63	17

Twice Measured Males (photographs taken in 1938–39
by D. L. Oliver and in 1967 by J. S. Friedlaender)

SIWAI

27–28   Tokura (at 14 and 42 years of age)

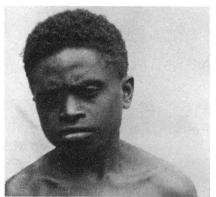 

29–30   Sipisong (at 15 and 43 years of age)

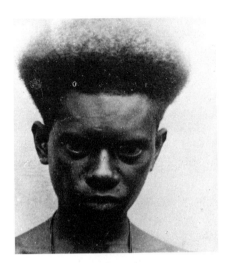 

31–32   Tumeriku (at 19 and 47 years of age)

 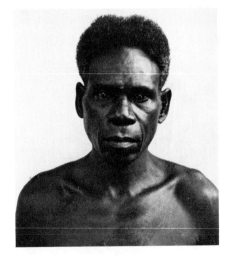

33–34   Tumare (at 20 and 48 years of age)

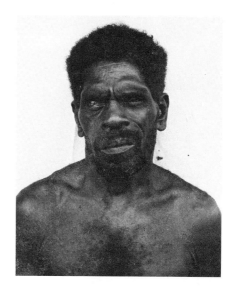

35–36   Sinkai (at 24 and 52 years of age)

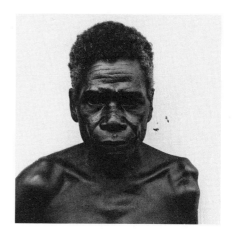

37–38   Pinoko (at 24 and 52 years of age)

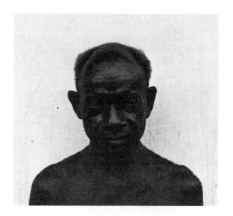

39–40   Tamang (at 25 and 53 years of age)

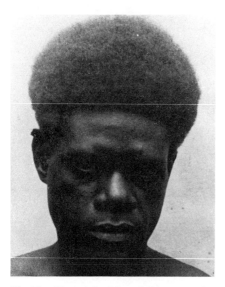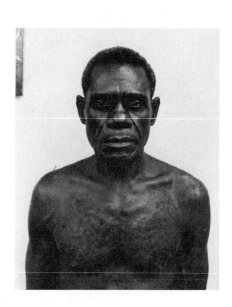

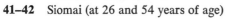

41–42   Siomai (at 26 and 54 years of age)

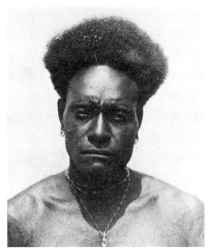 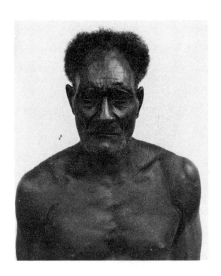

43–44 Napuai (at 30 and 58 years of age)

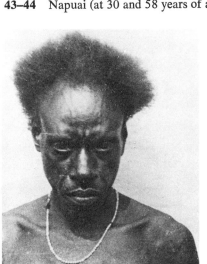 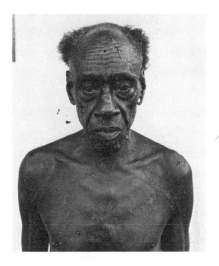

45–46 Kaijum (at 33 and 61 years of age)

 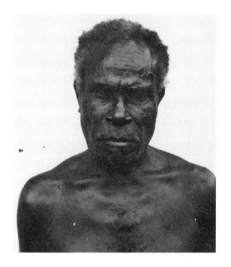

47–48   Kakai (at 34 and 62 years of age)

 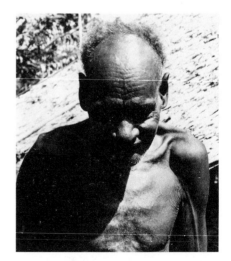

49–50   Tukeng (at 39 and 67 years of age)

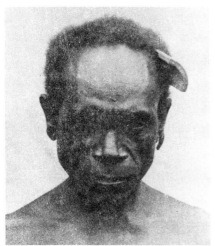 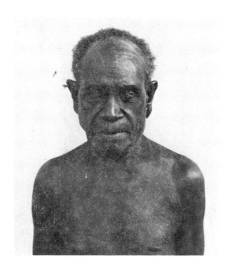

51–52   Kiwo (at 42 and 70 years of age)

TORAU

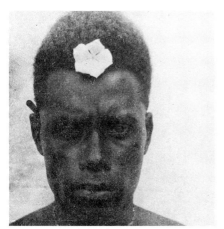 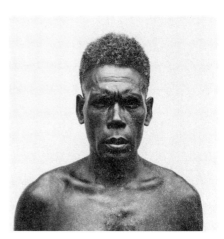

53–54   Malanu (at 22 and 50 years of age)

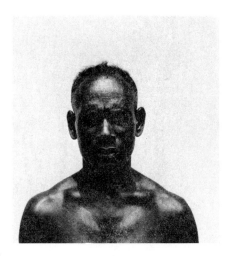

55–56 Tiotorau (at 23 and 51 years of age)

57–58 Aliali (at 24 and 52 years of age)

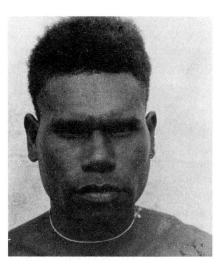 

59–60   Labo (at 26 and 54 years of age)

 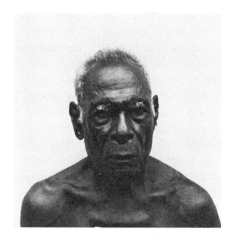

61–62   Gausu (at 55 and 83 years of age)

1938–39. This switch in survival differential set in earlier in the hard-pressed Siwai than among the eastern beach people, who generally seem to have enjoyed lower mortality rates. Perhaps bigger adolescents and young men survived the wartime deprivations more successfully, but die at increasingly high rates after the onset of middle age.

Univariate Analysis

The means and standard deviations of each measurement were calculated for each village sample and are listed in Tables 5.11 and 5.12.[3] Also, the F-ratios resulting from the analyses of variance and their significant levels are presented in Table 5.13. An analysis of variance measures the degree of heterogeneity between groups as opposed to the variability within each group for each measurement. All 13 variables show highly significant differences among the 18 samples, which come from a geographical range of no more than 50 miles. However, it is also clear from Table 5.13 that five of these measurements—head breadth, head length, chest breadth, nose height, and face height—have interpopulation differences which are much more important than any of the rest. When we look at the mean fluctuations on Table 5.11 with this in mind, two or three apparent clusters appear. Except for group 18—the Torau village of Rorovana—there seems to be a fairly definite break in absolute head breadth between the wide-headed northerners and the narrow-headed southerners. The Simeku villages (8–10), the transitional group between the southern Nasioi and the northern Eivo, who are classified linguistically as a "sublanguage" of Nasioi, fall in the northern range of means in head breadth. The same holds true for head length. The southern group, however, except for one village, all have means exceeding 190 mm. As for chest breadth, the Simeku means are more intermediate in position between the broad Eivo and Rotokas and the narrow Nasioi and Siwai. The Melanesian-speaking groups, 17 and 18, are in this case more similar to the northerners than to the southerners. As for nose height and face height, the Melanesian villagers fall together with the extreme northerners, the Rotokas and Aita (1 and 2), in having the longest faces. The

3. These results were calculated using the Data-Text series of packaged programs (Armor and Couch 1971).

Table 5.11
Anthropometric Group Means[a]

Village	Height	Arm lgth.[b]	Sit. ht.	Chest brdth.	Weight	Head lgth.	Head brdth.	Face ht.	Bizy.	Nose ht.	Bigon.	Min. front	Nose brdth.	N
1. Nupatoro	1651	772	851	251	130	187	147	120	139	52	106	103	43	21
2. Okowopaia	1640	758	852	249	130	186	149	119	140	52	104	103	45	19
3. Kopani	1636	772	843	265	132	186	148	117	141	51	104	104	44	54
4. Kopikiri	1613	754	837	261	127	181	148	112	139	47	102	102	44	10
5. Nasiwoiwa	1638	759	848	260	129	183	147	112	140	48	103	104	44	36
6. Atamo	1640	771	849	252	130	183	148	114	141	48	105	102	43	32
7. Uruto	1634	752	841	259	129	187	145	115	139	49	101	100	42	30
8. Karnavito	1604	771	837	251	127	185	146	113	140	50	105	101	43	35
9. Boira	1567	726	832	246	122	187	140	113	138	49	101	100	43	25
10. Korpei	1584	744	822	247	121	186	144	111	139	48	101	101	42	50
11. Sieronji	1604	749	830	244	124	191	139	114	137	48	103	101	41	9
12. Pomaua	1582	743	824	241	119	190	140	117	136	50	103	102	43	34
13. Bairima	1622	759	838	246	120	193	139	115	135	50	101	102	41	21
14. Turungum	1613	762	818	243	121	187	141	114	137	51	101	100	45	19
15. Moronei	1626	761	834	245	124	190	141	113	137	48	101	102	44	32
16. Old Siwai	1618	748	833	242	118	191	140	110	135	49	100	101	45	25
17. Arawa	1601	752	836	252	124	193	140	117	137	52	102	103	44	20
18. Rorovana	1635	767	853	256	132	193	144	122	139	54	104	103	43	54
														526

[a]All measurements are in millimeters except for weight, which is in pounds. The measurements are ordered from left to right by decreasing variance.
[b]Used in this and following tables to mean upper extremity length.

Table 5.12

Standard Deviations of the Group Anthropometric Means[a]

Village	Height	Arm lgth.	Sit. ht.	Chest brdth.	Weight	Head lgth.	Head brdth.	Face ht.	Bizy.	Nose ht.	Bigon.	Min. front	Nose brdth.	N
1. Nupatoro	38.00	30.98	21.15	10.41	10.22	4.62	3.57	6.93	4.21	*4.82	5.71	4.20	2.57	21
2. Okowopaia	51.74	32.25	28.00	9.65	13.98	5.48	2.79	4.10	5.82	2.29	5.06	3.50	3.20	19
3. Kopani	45.88	26.08	27.62	11.93	10.41	6.03	4.68	5.53	3.78	4.19	5.40	3.47	3.24	54
4. Kopikiri	30.60[b]	22.79	19.80	6.50	7.17	4.50	4.32	5.72	3.98	3.30	2.56	3.06	3.41	10
5. Nasiwoiwa	58.11	30.83	35.78	11.35	15.58	6.82	5.06	5.52	5.05	3.57	5.27	4.41	2.58	36
6. Atamo	65.17	33.30	27.27	9.46	11.68	6.50	5.26	7.23	4.91	3.30	5.47	4.46	3.44	32
7. Uruto	67.38	34.78	33.27	12.31	15.54	6.06	6.31	8.49	5.97	3.50	5.62	4.66	2.78	30
8. Karnavito	56.15	28.60	33.31	11.43	16.60	5.69	*6.61	5.52	5.23	2.84	5.46	4.21	3.40	35
9. Boira	70.46	35.82	30.24	9.00	10.96	5.35	3.76	6.62	4.45	3.52	5.55	3.77	1.62	25
10. Korpei	52.44	29.15	28.27	11.74	11.59	6.39	5.19	5.09	4.30	2.76	4.73	3.73	2.83	50
11. Sieronji	66.34	20.77	*46.70	11.15	15.32	*7.33	3.44	7.52	*6.23	3.71	6.62	4.92	3.16	9
12. Pomaua	59.24	32.58	24.63	11.50	12.18	6.43	4.82	6.62	3.91	3.62	*7.06	3.79	1.54	34
13. Bairima	64.54	*41.43	26.90	*15.79	*18.02	4.93	3.60	5.98	4.46	2.66	5.37	4.23	2.54	21
14. Turungum	48.51	39.12	24.55	10.88	12.71	4.74	3.39	7.23	3.09	3.60	5.15	3.42	3.05	19
15. Moronei	69.01	39.07	43.77	12.81	13.79	6.36	5.52	5.43	4.05	3.13	4.34	4.23	2.79	32
16. Old Siwai	*73.31[c]	38.64	33.92	13.46	15.19	5.48	6.18	*7.68	5.55	2.89	6.32	*5.74	*3.72	25
17. Arawa	44.87	26.84	22.83	12.63	14.72	5.85	4.39	4.97	4.65	3.39	7.03	4.35	3.12	20
18. Rorovana	55.05	34.11	32.21	13.23	16.26	5.71	4.43	6.23	5.56	4.02	5.98	4.04	2.98	54
Total	61.55	34.03	31.98	13.62	14.19	6.85	5.82	6.88	5.03	3.92	5.74	4.24	3.04	526

[a] All measurements are in mm. except for weight, which is in pounds. The measurements are ordered from left to right by decreasing variance.
[b] Underlined values are the smallest for each measurement.
[c] Asterisked values are the largest for each measurement.

Table 5.13
F-ratios and Their Associated Probabilities of Chance
Occurrence for the 13 Anthropometric Variables

Variable	F ratio	Probability
Head breadth	13.10	$.55 \times 10^{-27}$
Head length	11.28	$.45 \times 10^{-23}$
Chest breadth	11.21	$.63 \times 10^{-23}$
Nose height	9.16	$.27 \times 10^{-18}$
Total face height	7.77	$.47 \times 10^{-15}$
Stature	5.10	$.13 \times 10^{-8}$
Bizygomatic	4.58	$.25 \times 10^{-7}$
Arm length	3.74	$.30 \times 10^{-5}$
Nose breadth	3.65	$.48 \times 10^{-5}$
Sitting height	3.51	$.11 \times 10^{-4}$
Weight	3.38	$.22 \times 10^{-4}$
Bigonial	2.81	$.17 \times 10^{-3}$
Minimum frontal	2.79	$.19 \times 10^{-3}$

northerners as a group appear to be heavier, wider (both in the chest and head) and shorter-headed than the southerners. Within these gross divisions, however, it is hard to see any subclusters taking shape.

Correlation Coefficients and Analysis of Variance

Multivariate analyses require information on the associations of the variable used. The complete correlations of the 13 measurements over the entire sample are presented in Table 5.14. Other studies of the product moment correlation coefficients of measurements taken on the living, such as Majumdar and Rao (1960), Howells (1951), and Chai (1967), as well as this study, show that there are some small differences in the correlation coefficients derived from different sets of data, which usually amount to no more than a difference of .10. The heterogeneity of coefficients of correlation of the same measures in different groups in the Bougainville sample, and also between groups in the Indian sample, is probably best attributed to the small sample sizes in each case.

The table of correlation coefficients (Table 5.14) shows that body measurements, or at least the ones included in this test battery, are correlated at a

Table 5.14
Coefficients of Correlation for 13 Anthropometric Variables

	1	2	3	4	5	6	7	8	9	10	11	12	13
1. Weight	1.00												
2. Stature	.70	1.00											
3. Chest breadth	.70	.51	1.00										
4. Sitting height	.69	.79	.50	1.00									
5. Arm length	.62	.84	.47	.55	1.00								
6. Head length	.28	.26	.18	.27	.23	1.00							
7. Head breadth	.34	.25	.37	.19	.20	-.11	1.00						
8. Minimum frontal	.39	.29	.31	.22	.27	.26	.42	1.00					
9. Bizygomatic	.47	.33	.43	.29	.32	.07	.56	.48	1.00				
10. Bigonial	.43	.29	.30	.31	.26	.09	.31	.26	.46	1.00			
11. Face height	.31	.32	.26	.30	.31	.32	.16	.25	.20	.15	1.00		
12. Nose height	.10	.16	.13	.10	.21	.23	.11	.13	.15	.10	.59	1.00	
13. Nose breadth	.26	.15	.17	.16	.13	.10	.16	.15	.23	.16	.04	.14	1.00

much higher and more significant level than those taken on the head. Within the body measurements, the least highly correlated variable is chest breadth, which also has the highest F-ratio of any measurement on the body in this study. Weight, and then height, are the most strongly correlated individual measurements overall, and therefore the most redundant. One or the other, or both, are involved in the six strongest correlation coefficients, all with body measurements, especially with chest breadth and weight. Body breadths seem to vary with facial breadths. Nose breadth, the only measurement included which is taken on the soft tissue of the head, has the weakest correlations of all, followed by head length, a most important variable as judged by its F-ratio. Head breadth and head length, the two variables with the highest F-ratios, are the only two variables which are negatively intercorrelated, although very weakly. Nose height and face height, two of the other variables with high F-ratios, are very strongly correlated for head measurements, as one would expect. Nose height is the more independent of the two in relation to the other variables. It seems clear that nose height and face height are highly redundant, while the other variables with high F-ratios, head length, head breadth, and chest breadth, are all adding new information on morphological variation. These last three variables might appear in different or contrasting positions in multiple discriminant analysis.

Discriminant Analysis

The assumptions of discriminant analysis are essentially the same as those for the analysis of variance and are fulfilled by anthropometric variables, save possibly for weight. The inclusion of the two small samples in the present instance might be cause for concern, as mentioned above, because the group variance-covariance matrices (dispersions) might have proved to be heterogeneous. The H_1 test of homogeneity of variance in all 18 groups[4] shown below (Table 5.15) indicates that there is very little difference in group dispersions except for the two small samples, which cause the total F-ratio to be slightly significant. However, this significance is due principally to the large

---

4. The H_1 null hypothesis asserts that the sample populations have equal multivariate dispersions. For an extended discussion, see Cooley and Lohnes (1962:60–63).

Table 5.15
H_1 Test for the Homogeniety of Group Dispersions

Number	Village	Log determinant
1.	Nupatoro	49.0
2.	Okowopaia	46.3
3.	Kopani	48.9
4.	Kopikiri	14.6
5.	Nasiwoiwa	51.5
6.	Atamo	49.0
7.	Uruto	47.1
8.	Karnavito	47.0
9.	Boira	44.1
10.	Korpei	47.5
11.	Sieronji	12.0
12.	Pomaua	44.6
13.	Bairima	43.1
14.	Turungum	49.6
15.	Moronei	51.5
16.	Old Siwai	54.6
17.	Arawa	48.9
18.	Rorovana	54.1

[a]DF_1 = 1547; DF_2 = 34376

For test of H_1, F = 1.654, significant at the .01 level.

[a]DF = degrees of freedom.

values attained by the second degree of freedom and should not affect the discriminant test.[5]

The means of calculating the linear discriminant function is by axis rotation, somewhat like factor analysis. However, instead of simply trying to attain a more elegant description of the variation within a single sample as in factor analysis, here the goal is to create a vector or axis which will discriminate the known groups better than any single given variable. This rotation is performed on the unstandardized, or raw, scores by multiplying them by the discriminant

5. Personal communication of Dr. Kenneth Jones.

vector weights, which act as cosine coefficients of the angle through which the axis is rotated. A typical equation therefore takes the form:

$$v_1 x_1 + v_2 x_2 + \ldots v_n x_n = d,$$

where "$x_1 \ldots {}_n$" are the variables and "$v_1 \ldots {}_n$" are their "weights," or coefficients. Finding the vectors involves maximizing the ratio of the among-groups sum of squares to the within-groups sum of squares. For two groups there is only one discriminant axis possible, no matter how many variables are used. For the general case, the number of functions is limited to one less than the number of groups or variables, whichever is smaller. Usually only a few of these multiple discriminants are significant.[6]

The significance of the entire discrimination is indicated by Wilks's Lambda test, which asserts as its null hypothesis the equality of the population centroids (the test of H_2) assuming H_1. Rao (1963) defines Wilks's Λ criterion as follows: $\mathbf{W}/\mathbf{T}$ or, more properly, $\mathbf{W}\mathbf{T}^{-1}$, where $\mathbf{W}$ is the pooled within-groups deviation score cross-products matrix and $\mathbf{T}$ is the total sample deviation score cross-products matrix. As $\mathbf{T}$ increases relative to $\mathbf{W}$, the ratio decreases in size with an accompanying increase in the confidence with which H_2 can be rejected. As a test of significance for Λ, Rao's F is considered most appropriate. The results indicate highly significant discrimination overall (see Table 5.16). The possibility that the population centroids are not significantly different can be rejected at the .001 level.

The significances of the 12 individual functions (there being 13 variables) and their relative importance are also presented in Table 5.16. The first four, or possibly the first five, discriminants, are the only important axes and account for 90 percent of all discrimination. Only these five "dimensions" were used in the further calculations of group means and individual scores in the multidimensional space.

It is important to understand what the individual functions accomplish and which variables are stressed. Two sets of coefficients have been used in previous analyses, the scaled vectors and discriminant loadings. Scaled vectors are derived by multiplying the raw vectors by the square root of the pooled within-

6. For a more detailed description see Appendix.

groups deviations for that variable, thereby standardizing the values by the variances of the variables. These values are independent of one another and represent only the unique contribution of a variable. Interpretation of the nature of the discriminant equations from their scaled vectors can often be misleading and difficult as a result.

The other diagnostic coefficients, the discriminant loadings, are more directly interpretable. These are analogous to factor loadings in that they are the actual correlations between the discriminant scores and the scores for each variable. One drawback of the discriminant loadings is that a measurement which has very little to do with distinguishing members from different groups may have a large discriminant loading because it is highly correlated with a variable which is a strong discriminator. In this sense, discriminant loadings may be misleading. Nevertheless, I use them here because they are more directly interpretable. If all the measurements are standardized and uncorrelated, then the vectors, scaled vectors, and discriminant loadings will all take on the same values.

The first function, it is clear from Table 5.17, acts to separate the northern

Table 5.16
Relative Importance of Male Anthropometric Discriminant Functions

Wilks's Lambda for total discrimination = .156
Rao's F = 4.591
DF_1 = 221 DF_2 = 5180.7
Chance probability of F = .586 x 10^{-14}
Decreasing probability of significance of successive roots

Root	Eigenvalue	Percent of trace	Chi-square tests of significance with successive roots removed χ^2	Probability
0			946.2	1.000×10^{-38}
1	1.178	46.1	549.5	$.615 \times 10^{-24}$
2	.391	17.7	381.5	$.335 \times 10^{-13}$
3	.291	10.7	251.5	$.607 \times 10^{-4}$
4	.262	10.3	132.8	$.111 \times 10^{-1}$
5	.177	4.9	49.7	.230

[a]DF = degrees of freedom.

Papuan-speakers from the southern Papuan- and Melanesian-speakers. This is the brute discriminator, accounting for 46.13 percent of trace (Table 5.16). The four relatively "pure" Eivo villages (3–6) are the extremes on the upper end of the scale. Scoring lower are the two Rotokas and Aita villages (1 and 2), two closely related Simeku villages, 8 and 10, and the one Eivo village, 7, which has over .50 interbreeding frequencies with the Simeku. The other Simeku village, 9, falls on the other side of a large gap, in the upper limits of the large group of southern villages that range in discriminant scores from −5 to −1.6. According to the villagers themselves, village 9 was historically the most southerly of the Simeku villages I sampled. The Nasioi have the lowest scores of all, with the Melanesian-speakers and Siwai scoring somewhat higher.

From the list of discriminant loadings (Table 5.18), it becomes clear that the first discriminant function amounts to a refined Cranial Index with ramifications throughout the body. Head breadth and head length are both strongly correlated with the first function (length with a negative sign), and the other

Table 5.17
Male Anthropometrics: Group Centroids in Discriminant Space

Village	Function 1	Function 2	Function 3	Function 4	Function 5
1	.20	.75	.49	.72	−.03
2	.61	.38	.99	1.07	−.42
3	1.41	.57	−.23	−.33	.34
4	2.09	−.06	.17	−.64	.19
5	1.61	.18	.20	−.80	−.47
6	1.21	−.40	.31	.66	.02
7	.65	−.03	−.68	.07	.28
8	.45	−.53	−.25	.52	−.04
9	−.54	−.87	−.86	−.08	−.49
10	.13	−.93	−.27	.07	−.05
11	−1.43	−.40	−.22	.00	−.25
12	−1.50	−.25	−.10	.09	.11
13	−1.56	.07	.03	−.65	−.11
14	−.54	−.47	1.00	.14	.49
15	−.82	−.51	.81	−.31	.07
16	−1.12	.17	.87	−.73	−.07
17	−1.33	.62	−.34	−.60	.16
18	−.94	1.12	−.49	.42	.00

variables seem to be correlated with the discriminant more or less to the degree that they are correlated with head breadth. Chest breadth and bizygomatic breadth, two important variables in terms of their F-ratios, are strongly correlated with the first discriminant. These breadths are probably the principal modifiers which make the ordering of the villages on this discriminant slightly different from the simple ranking by cranial index scores presented in Table 5.19. Nor is there such a clear break into two groups in the C.I. results as in the discriminant scores.

The second discriminant function, which accounts for 17.66 percent of the total discrimination, serves to separate the Melanesian-speakers (17 and 18) from the rest of the southern villages. The Rotokas and Aita villages (1 and 2), along with the northernmost and southernmost Eivo villages, also have extreme values. This discriminant differentiates within the two large groups that are defined by the first function. From the discriminant loadings it appears that this function reflects changes in face height, nose height, and chest breadth, along with general bigness throughout the body. The three variables most strongly correlated with this function—nose height, face height, and chest

Table 5.18
Discriminant Loadings
(Correlation of each anthropometric variable with each discriminant function)

Variable	Function 1	Function 2	Function 3	Function 4
1. Weight	.17	.36	−.07	.19
2. Stature	.16	.41	.31	.03
3. Chest breadth	.40	.54	−.36	−.20
4. Sitting height	.09	.34	−.04	.09
5. Upper extremety	.12	.28	.16	.05
6. Head length	−.51	.37	−.01	.00
7. Head breadth	.53	.25	.14	.45
8. Minimum frontal breadth	.08	.37	.11	−.13
9. Bizygomatic	.31	.09	−.07	.30
10. Bigonial	.11	.19	.00	.34
11. Face height	−.10	.67	−.10	.44
12. Nose height	−.13	.41	.11	.00
13. Nose breadth	.11	.09	.42	−.02

breadth—have the highest F-ratios of all measurements except head length and breadth. Those villages which score highest have the biggest men.

Function 3, accounting for 10.73 percent of the total discrimination, distinguishes other subgroups. At the positive end of the scale are the Rotokas and Aita once again (1 and 2), and the Siwai (14–16), who are clearly distinguished on this axis from the Nasioi (11–13) and Simeku (8–10). Through the third discriminant function, villages belonging to the same linguistic group are never greatly separated.

The weighting of the variables in function 3 is rather peculiar. Height and nose breadth are contrasted with chest breadth. The first discriminant or two ordinarily account for the principal morphological relationships which differentiate the samples, and the remaining discrimination is described by functions with little apparent biological coherence. For, while each new discriminant is independent of the preceding ones, different morphological relationships are

Table 5.19
Cranial Index for the Bougainville Series

Village	Cranial Index
1. Nupatoro	78.6
2. Okowopaia	80.6
3. Kopani	80.1
4. Kopikiri	82.3
5. Nasiwoiwa	80.3
6. Atamo	81.4
7. Uruto	77.5
8. Karnavito	78.5
9. Boira	75.8
10. Korpei	78.0
11. Sieronji	72.8
12. Pomaua	73.3
13. Bairima	72.5
14. Turungum	75.5
15. Moronei	73.8
16. Old Siwai	73.8
17. Arawa	72.7
18. Rorovana	74.2

usually interrelated. Discrimination has the sole aim of group separation, not biological relevance.

Functions 4 and 5 do seem to break up the homogeneity of the different linguistic units, but of course account for much less of the total discrimination.

The various effects of Functions 1 through 3 are more comprehensible when the centroids of each group are plotted in space defined by the functions three at a time (see Figure 5.9). The graphs reveal the overriding importance of discriminant function 1, where the group means are spread over a very wide range. This reflects the large proportion of total discrimination which discriminant 1 accounts for. The large gap between the northern and southern groups along function 1 is quite striking, and another break between the core

5.9 Anthropometry: centroids of the 18 samples on the first three discriminant axes

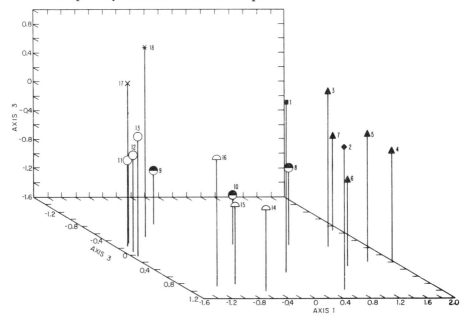

Eivo (3–6) and the other northerners is apparent enough. The figure reveals the closeness of the separate linguistic clusterings along all three axes.

A last step in discriminant analysis, which is perhaps the best way of establishing the significance of the separateness of the different groups (as opposed to the calculation of distances between their centroids), is the calculation of a "hits and misses" table such as Table 5.20. The actual group memberships of the individuals in the sample are plotted against their predicted group memberships, and these scores are converted into percentages. Village by village, the following features stand out. Although village clusters overlap a great deal, there is no question that the total discrimination is highly significant. The average percentage of hits ranges from a low of .17 for centrally placed Karnavito (8) to .90 for Kopikiri (4) at the extreme on function 1. However, two of the three very high scorers, groups 4 and 11, are the two smallest samples. The other high-scoring village, Okowopaia (2), is evidently the most isolated and divergent village in the entire survey. Generally, most villages have a correct assignation value of about .30. Table 5.20 is easily divisible into four sections when separating vertical and horizontal lines are drawn between villages 10 and 11. The upper left and lower right sections of the table have the great preponderance of assignations, while the two other sections are sparsely populated. There are no percentages above .07 in these two sections except for the two reciprocal combinations of groups 17 and 9. This gross division corresponds to the linguistic separation between the northern Papuan groups and the rest, except that the Simeku group 8–10, falls with the northerners. The first discriminant function reflects the overriding importance of this general pattern.

The role of the language group as morphological unit becomes more apparent when the village assignment scores are converted to language group percentages (Table 5.21). In this table, the Arawas, group 17, are "closer" in percentage scores to the Nasioi, Simeku, and Siwai than they are with their old linguistic cousins in Rorovana (18), who, on their part, have affinities with the Simeku, Eivo, and Nasioi as well as the Arawas. This seems logical in light of the interbreeding frequencies given in Table 3.8, where the Arawas are seen to be interbreeding with the Nasioi at a rather high rate, and with village 13 in particular. The Rorovanas are further north, closer to the Simeku and Eivo.

Table 5.20
"Hits" and "Misses"[a]
(Predicted village membership as columns, actual village membership as rows)

Village	1	2	3	4	5	6	7	8	9	10	11	12	13	14	15	16	17	18	N
1.	.333[b]	.095	.095		.095	.048	.048	.095			.048			.048	.048			.048	21
2.		.895						.053						.053					19
3.	.074	.037	.315	.130	.148	.056	.074	.019		.037					.019		.019	.056	56
4.				.900	.100														10
5.	.028	.028	.167	.167	.389	.056	.028	.028	.094	.056					.028	.028			36
6.	.031	.031		.063	.031	.344	.063	.156	.033	.063				.031	.063	.031			32
7.	.100	.033	.033	.033		.067	.433	.033	.143	.033	.067		.033		.033			.033	30
8.	.057	.086	.029	.057	.086	.029	.086	.171		.086				.057			.029	.057	35
9.									.520	.080	.080						.120		25
10.	.020	.040	.040	.040	.040	.040	.100	.060	.080	.320	.040	.040	.040	.020	.040	.020	.020		50
11.											.889								9
12.								.059	.059		.206	.206	.088	.059	.029		.148	.088	34
13.											.143		.429		.048	.095	.190	.048	21
14.						.053	.053			.053	.053		.053	.368	.105	.053	.053	.053	19
15.				.031		.031	.031	.031		.062	.031	.094	.094	.186	.219	.094	.031	.031	32
16.									.080		.080	.040	.120	.160	.080	.280	.080		25
17.									.150		.100		.200	.050		.050	.400	.050	20
18.	.074	.018	.037	.018	.018		.074	.018	.018		.074	.018	.018	.056	.018	.018	.111	.407	54

[a]Total percentage of "hits" = 37.6%, significant at the .001 level.
[b]The diagonal elements ("hits") are underscored.

Table 5.21
"Hits" and "Misses" Table Grouped by
Linguistic Affiliation[a]

	Aita	Rotokas	Eivo	Simeku	Nasioi	Siwai	Uruava	Torau	N
				A. Absolute Numbers					
Aita	7[b]	2	6	2	1	2		1	21
Rotokas		17		1		1			19
Eivo	9	5	111	19	4	9	1	4	162
Simeku	3	5	25	54	10	6	5	2	110
Nasioi			1	6	38	6	9	4	64
Siwai	1		8	7	15	39	4	2	76
Uruava				3	6	2	8	1	20
Torau	4	1	8	2	6	5	6	22	54
									526
				B. Converted to percentages					
Aita	.333	.095	.286	.095	.048	.095		.048	
Rotokas		.895		.053		.053			
Eivo	.056	.031	.685	.117	.025	.056	.006	.025	
Simeku	.027	.045	.227	.491	.091	.055	.045	.018	
Nasioi			.016	.094	.594	.094	.141	.063	
Siwai	.013		.105	.092	.197	.513	.053	.026	
Uruava				.150	.300	.100	.400	.050	
Torau	.074	.019	.149	.370	.111	.093	.111	.407	

[a]Percentage of "hits" now equals 56.2%.
[b]The diagonal elements ("hits") are underscored.

The Eivo, Rotokas, and Aita, all belonging to one language family, have the highest rates of correct assignation and the lowest rates of interbreeding, dwelling in the most remote villages in the most sparsely populated area in the survey. So there is a clear negative correlation between interbreeding frequencies and correct assignation by anthropometric description. Note in particular that the mixed group, the Simeku, share higher interbreeding and assignment frequencies with the Eivo than with the Nasioi, with whom they are classified linguistically.

Conclusion

This multivariate analysis indicates that the most important of the measurements which I took for distinguishing between groups are those connected closely with bone growth. In comparing these results with those of Hiernaux's study (1963) of heritability in highland and lowland Hutu, clearly those variables which differentiate the Hutu groups in different ecological zones are not important in distinguishing the Bougainville populations which I studied. The anthropometric differences which Hiernaux found were principally those describing soft tissue and absolute size. The relation of head length to breadth, so important in the Bougainville series, is constant in Hiernaux's two groups, although the absolute size of both measurements do change from one to the other. Weight and nose breadth, other measurements which Hiernaux finds to vary significantly between his groups, are relatively stable in the Bougainville series. The most likely explanation is that the morphological differences encountered in the present study are genetically rather than environmentally determined. This also largely corresponds with Osborne and DeGeorge's (1959) conclusions concerning the heritability of different anthropometric traits.

As for the pattern the villages make in the discriminant analysis, it is a neat reflection of language group discreteness, with consistent clustering of villages from the same language area. The actual relationships of groups from different language areas also resemble the language pattern in a general way. It appears to be a more distinct pattern, in this respect, than the more purely "genetic" map of blood relationships, and bears little resemblance to the patterns predicted from current migration rates. Language groups which share comparatively high intermarriage frequencies tend to intergrade with one another morphologically, as one would expect from extensive gene flow. Examples are the "poly-clines" between the Eivo-Simeku-Nasioi, between the Torau and Nasioi, and even within the separate language groups. Broader relationships between language units which are closely related but no longer have high rates of marriage exchange, such as the Siwai and Nasioi, or the Rotokas, Eivo, and Aita, do occupy neighboring positions in morphological space, but do not form intergrading clines.

In sum, there is strong circumstantial evidence that the differences emphasized by the discriminant analysis of the anthropometric variables are genetically determined ones, and that the likely cause for the pattern is to be found in historical relationships of migration and marital exchange rather than in selective differences among the groups. This is in spite of the clear evidence for a secular trend in weight and height among the Siwai and Rorovana, and presumably the other more acculturated groups in the survey, which might otherwise obscure these genetic relationships.

6 Finger and Hand Prints

Finger and hand prints have fascinated human geneticists because the basic patterns are set down quite early in fetal life (by the third month) and are essentially unchangeable thereafter. Everyone knows that each individual, and in fact each finger of each individual, has unique ridge configurations. Yet the grosser patterns of the fingertips and palms show a great many similarities among the fingers of a given individual, among the corresponding fingers of related individuals, and, to a far lesser extent, among individuals from the same population. All this implies strong genetic determination or control of these gross patterns, unencumbered by any environmental influences. Since ridges are countable, repetitive features, there is the added lure of quantification, always an advantage in genetics.

The fingertips have the simplest ridge patterns to analyze. Simplest of all are those on which the ridges run more or less parallel to one another across the tip in an "arch" (see Figure 6.1). The second type of pattern is characterized by a looping arrangement, where some ridges begin near the flexion creases on one side, curve up and around the ball of the finger, and loop back on themselves. For such patterns, there is usually one spot on the opposite side of the tip from the entry and exit points of the loop where ridges converge from three different directions. Such a "triradius" is missing in true arches. Loops may be subclassified into those which open to the outside, or ulnar, side of the hand, and those which open to the radial, or thumb, side. A third major pattern type is the whorl, where there are ordinarily two triradii at either corner of the base of the tip and a circular arrangement of ridges in the ball of the tip. There are a number of intermediate patterns, such as central pocket loops and compound prints, but these are beyond our discussion (see Holt 1968, Cummins and Midlo 1961).

The traditional way of quantifying ridge pattern variation in a gross fashion is to count the number of ridges between each triradius and the center of the loop, or whorl, of that finger (obviously, for an arch, the count is 0). This is an extraordinarily tedious process which has to be done by human hands and eyes. If there are two triradii on a finger (as in a whorl), the largest ridge count may be taken to represent that finger (a "maximum" count), or both counts may be summed, for a "total" count. The Total Ridge Count is the sum of the maximum counts for the 10 fingers.

6.1   Classification of dermatoglyphic patterns

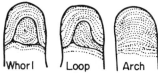

DERMATOGLYPHICS

Whorl   Loop   Arch

Three basic types of finger prints: whorl, loop, arch.

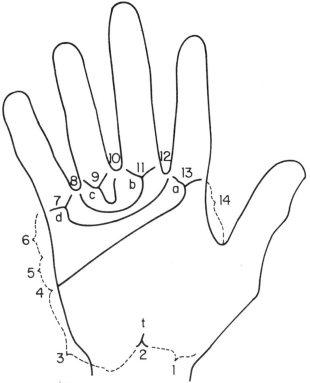

The scheme of numbers for formulating palmar main lines.

On the palm, triradii generally appear at the base of each of the four fingers except for the thumb, and also somewhere at the base of the carpals, usually over the carpal tunnel. They and their emanating lines have been designated A, B, C, D, and T (see Figure 6.1), and the patterns these lines form on the palm have been the objects of some more limited studies (Holt 1968, Pons 1959). At least for mainlines A and D, the higher the "score," the more transverse the arrangement of ridges, while low values are associated with longitudinal alignment.[1] There are many other dermatoglyphic features which have been studied elsewhere (e.g., the *atd* angle, the presence and nature of auxiliary triradii, and complex patterns on the palm), but these seem to be less closely tied to the individual's genotype, and in the present survey showed little differentiation among groups.

Heritability Estimates of Ridge Patterns

As implied above, fingerprints have been singled out as reflecting the least distorted products of polygenetic variation that have an approximately normal distribution. In the language of quantitative genetics, finger ridge counts have very high heritabilities. This means that, in a simplified way, the variation in ridge counts from one individual to the next is caused by differences in the genes of those individuals rather than by differences in their environments or upbringing. "Heritability" refers to the proportion of the variation in a characteristic within a population which is determined by genes rather than by environmental "noise." However, it is a common statement in quantitative genetics that an estimate of heritability only holds for a particular population in a particular environment. Changing the genetic composition of the population or changing the nature of the environment will almost inevitably change the relative contributions to physical variation of the environment and the genotype.

1. The scheme of palmar scoring given in Figure 6.1 is a modified version of Cummins and Midlo (1961). Here, to standardize scales, this scheme was used only for mainline A. For mainline B, the points were renumbered as follows:
$< 5 = 1, 5 = 2, 6 = 3, 7 = 4, 8 = 5, 9 = 6, > 9 = 7$.
For mainline C:
$0 = 0, < 7 = 1, 7 = 2, 8 = 3, 9 = 4, 10 = 5, 11 = 6, > 11 = 7$.
For mainline D:
$< 8 = 1, 8 = 2, 9 = 3, 10 = 4, 11 = 5, 12 = 6, 13 = 7, > 13 = 8$.

For a continuously varying character, a perfect heritability score of 1.00 means that the proportion of the genes any two relatives hold in common should reflect exactly the proportional reduction in variation, or elevation in similarity, of those two people, when compared with variation among randomly chosen individuals from that population. Table 6.1 gives the expected correlations under such ideal circumstances. Identical, or monozygotic, twins, with identical sets of genes, should have a correlation coefficient of 1.0.

The results from Holt's 1968 dermatoglyphic study, in particular, are in near perfect agreement with the expectation that there is almost no environmental contribution to variation in ridge counts, and that the genes which do affect the configuration of the patterns act in an equal, or additive and nondominant, fashion. Holt's results have been widely accepted (see Cavalli-Sforza and Bodmer 1971, Stern 1960, Lerner 1969), but, they are not without their problems. First of all, her work has been criticized by Froehlich (1974) and others (Weninger 1964; Parsi and DiBacco 1968) on the grounds that her sample of 100 families from Britain included 21 Welsh and 10 Jewish families, which would tend to exaggerate the differences among family groups and inflate the resulting correlation coefficients. In the same vein, Howells (1966b) showed that higher heritability estimates in anthropometric results came from "general" U.S. population studies than from a study of a religious isolate (the Hutterites). He attributed this difference to the greater genetic and environmental heterogeneity among families from the general population, while variability within family lines was essentially the same in both populations. This served to inflate differences in the broader survey over what they would have been if the population were a homogeneous one, as most genetic tests assume.

In fact, for studies on the heritability of dermatoglyphic variables from smaller isolated populations (e.g., Loesch 1971 on rural Poles, Hunt et al. 1964, on Yapese, and Froehlich 1975 on Solomon Islanders) estimates have run a good deal lower than Holt's. In particular, Froehlich, in his massive study, found that variation in ridge count for each finger, as well as summed over all fingers, is 70 percent attributable to genetic variation, at least in his two samples from Malaita Island and from the Aita in north Bougainville.

A major problem in such comparisons lies in the weakness of the correlation technique in defining the level of genetic determination. On Fisher's theory

Table 6.1
Ideal Coefficients of Correlation among Relatives
with Regard to Genetically Determined Characters

Relationship	r
Identical twins	1.00
Fraternal twins	.50
Parent-offspring	.50[a]
Siblings	.50
Uncle-nephew	.25
First cousins	.125

[a]Can be affected by sex.

(the source for Table 6.1), the parent-offspring correlation is doubled to estimate the additive genetic proportion of the total variance. A correlation of .50 means that a trait is totally determined by several nondominant additive genes and has a heritability of 1.0. But when such an intraclass correlation is doubled, so are its confidence limits, so the estimate of heritability is even less precise. For Froehlich's data on the Aita in north Bougainville, the sib correlation for the total ridge count is .416 and the confidence interval within which the true value is to be found is .30 to .52. Doubling both, the estimate of genetic determination for this trait is about .83 and the confidence interval is from .60 to 1.00!

In an effort to reduce the confidence limits of his various estimates, Froehlich combined ridge counts from each finger after the correlations had been calculated. This averaging of the individual correlations does not inflate the estimate of genetic variance, as seems to happen with the total ridge count. But, most important, the average correlation has the advantage of a reduced standard error, making statistical comparisons and estimates of genetic determination more precise.[2] The resulting genetic correlations for "digital pattern size" was

2. As Froehlich points out, correlation coefficients can only be averaged in this fashion if they appear to come from the same statistical universe. From both of Froehlich's populations, a chi-square test of homogeneity was applied to the ten sets of sibling ridge counts, and the results were nowhere close to significance (P > .9 in every case), indicating a high degree of homogeneity, at least.

.329 ± .017 for the Aita and .361 ± .022 for the Baegu of Malaita. These results are not significantly different from the total ridge-count correlations of their respective populations. They suggest that a realistic estimate of the genetic determination of the variation in ridge counts is in the vicinity of .70, with a confidence range of less than .60 to .80.

As for other dermatoglyphic traits, in Froehlich's survey they show approximately the same order of genetic determination, or somewhat less. Pattern intensity indices do not deviate significantly from the genetic correlations for total ridge count, indicating some comparable genetic control for the number of triradii. The intraclass correlations for palmar traits show far less genetic control than the digital variables. The total mainline index has the highest value, .307 ± .039 for the Aita. This suggests a proportion of genetic determination of about .50, much less than the value reported by Pons (1959). The individual mainlines have a genetic determination of about 35–55 percent. Interdigital patterns and hypothenar patterning are even less informative concerning genetic variation. For the two Solomon Island populations, no significant differences were found between results for the two populations, and in most cases the heritabilities were quite close.

The logical conclusion is that dermatoglyphic characters, and in particular finger ridge counts, should be sensitive indicators of genetic differences between populations. However, some earlier studies suggest a different outcome. Notably, Chai (1967) applied both multiple discriminant analysis and Mahalanobis' D^2 analysis to anthropometric and dermatoglyphic sets of data from different Taiwanese tribal samples. In Chai's survey, the discriminatory power of the anthropometric set of variables was much greater than the dermatoglyphic set, and the average D^2 was much larger for the anthropometrics. He used a set of 19 anthropometric measurements, more exhaustive than mine, but in the dermatoglyphic analysis did not use the time-consuming ridge-counting procedures. The same general differences in discriminatory power have been noted in other studies, notably in the Markham Valley of New Guinea (McHenry and Giles 1971; Froehlich, unpublished). Chai explains the difference by suggesting dermatoglyphic traits may be influenced by a smaller number of genes so that large sampling errors are compounded, and also by suggesting that finger patterns are less adaptively important than what is reflected in anthropometrics. He argues that the effects of natural selection may be greater on physical characters which he measured than on dermatoglyphics, and differential selection has been the causative factor in the variation among Taiwanese groups.

On the other hand, Froehlich thinks that the ridge patterns are influenced by a very large number of genes, that the actual patterns are a compromise among a large number of genetic influences, all of which come to bear at the end of the first trimester of uterine life. A common analogy of the determination of fingerprint patterns has been to the patterning of sand by the action of waves and wind (Stern 1960). This implies that during the short period of time in which the ridge configurations are determined, almost anything can influence them in some fashion. In genetic terms they are poorly canalized.

No matter the theoretical concerns, it is patently true that variation within a single population in ridge patterns is enormous—the within-group variances are quite large, while, comparatively speaking, the differences in the averages of different populations are quite small. This inevitably makes for smaller F-ratios and less significant discrimination among groups.

There is always the nagging suspicion that the conventional methods of analyzing dermatoglyphics are not only tedious but inadequate to the task, and that there should be ways of classifying variation in ridge patterns which are more direct reflections of development and genetic control. Some students in this field are currently attempting to apply trend surface and Fourier analysis to ridge patterns, and certainly these seem highly logical tacks to take. Factor analyses of ridge counts (see Froehlich 1974, Radzewi et al., unpublished) give different results, depending upon what is being included and how the ridges are counted.

Statistics

The dermatoglyphic sample from these Bougainville groups is the most exhaustive of any sample I have. Essentially everyone interviewed or counted in my census was inked and printed, even the youngest infants. Printer's ink on typing paper proved highly satisfactory and inexpensive and has not faded over time, as do some inkless methods. The arduous task of counting ridges and classifying patterns was performed by two undergraduates, Christopher Hitt and

Brian Eisenberg, who periodically checked each other's counting results. Those below ten years of age were excluded from the final analysis; the samples were considered of sufficient size without them, and since all the parents were included, we had an adequate representation of the various gene pools.

The male and female sets of prints were analyzed separately, and with good reason. First of all, the numbers in each group were considered adequate (being comparable in size to Holt's British survey), and, as will become apparent, the sexes differed appreciably in their mean values for different characters, and characters which vary widely in males among the different groups may be more uniform in females. As it turns out, the pattern of discrimination among the villages is quite different in the male and female series.

Table 6.2 presents the means and standard deviations for the mainline scores, maximum, and total ridge counts for males and females, in the total sample. These statistics are useful when the variables are normally distributed, and dermatoglyphic variables are not, strictly speaking. The salient points are: (1) males have consistently lower mainline scores than females; (2) males have larger values than women for the first and fifth fingers for both maximum and total ridge counts, while there are a number of reversals in the three middle fingers; (3) the standard deviations for males are slightly larger for the first and fifth fingers, while the variation for the middle three fingers is mixed; (4) the overall correspondence of means and standard deviations for the two sexes is considerable. Right mainlines have larger mean scores than left, and the lowest variation is in the fifth finger.

Correlation matrices of the different variables were calculated under the assumptions of normality. The resulting correlation matrices for the two sexes, given in Table 6.3, are again quite similar, with a few notable exceptions. The mainline scores are essentially independent from the finger ridge counts and are generally correlated at low levels with one another. Mainline D has the highest average correlations with the other mainlines, above all with mainline B from the same hand. The right hand shows higher intercorrelations than the left, so that in order of magnitude, the correlations for the right D mainline are with right B, right C, and then left D. For the left hand, the order is left D, left B, right B, and right D. That is, the right patterns are more tightly interrelated, both among themselves and to the left patterns. This implies stricter canaliza-

Table 6.2
Male Dermatoglyphics—Mainline Patterns

Group	A R	A L	B R	B L	C R	C L	D R	D L	N
1	4.77	3.95	3.86	3.23	3.18	1.95	4.68	3.64	22
2	4.13	2.96	3.79	2.75	2.54	2.13	4.42	3.50	24
3	4.86	3.61	4.05	3.22	3.47	2.27	4.86	4.03	59
4	5.05	3.91	3.95	3.41	3.55	3.05	4.41	4.00	22
5	4.67	3.67	4.31	3.13	3.49	2.47	5.02	3.91	45
6	4.52	3.25	4.00	3.29	3.46	2.69	4.67	3.81	52
7	4.76	3.73	4.64	3.36	4.21	2.61	5.12	4.09	33
8	4.45	3.17	3.66	2.98	3.47	2.64	4.32	3.79	47
9	4.87	3.97	4.58	3.13	4.26	2.42	5.23	3.90	31
10	4.58	3.32	3.79	3.14	3.08	2.24	4.59	3.88	66
11	4.62	3.62	4.77	3.54	3.62	2.08	5.23	4.38	13
12	4.58	3.38	4.00	3.32	3.18	2.36	4.68	4.12	50
13	4.95	3.64	4.45	3.55	4.23	2.95	5.41	4.50	22
14	4.86	3.29	4.38	3.33	3.38	2.29	5.19	4.00	21
15	4.76	3.21	4.73	3.15	4.58	2.67	5.48	3.88	33
16	5.00	3.33	4.50	3.17	4.46	2.04	5.00	3.96	24
17	4.42	2.54	3.46	2.73	3.19	1.85	4.08	3.15	26
18	4.56	3.05	3.54	2.88	2.95	1.92	4.14	3.42	85
Total	4.66	3.38	4.04	3.15	3.48	2.35	4.72	3.85	675
Std. Dev.	0.96	1.29	1.22	1.05	1.85	1.56	1.36	1.36	

Table 6.2 (continued)
Male Dermatoglyphics—Maximum Finger Ridge Counts

Group	T R	T L	2 R	2 L	3 R	3 L	4 R	4 L	5 R	5 L	N
1	13.91	9.82	10.77	9.23	12.55	12.14	16.27	16.05	12.73	11.91	22
2	20.13	17.04	15.08	15.33	13.58	15.50	18.42	19.13	12.88	14.96	24
3	17.51	15.32	12.88	13.80	16.71	17.05	20.22	19.98	13.88	13.63	59
4	16.27	15.59	12.09	10.77	13.73	13.00	17.23	17.68	13.32	11.86	22
5	19.67	17.78	16.36	15.80	17.00	17.13	21.13	20.51	15.29	15.22	45
6	19.60	15.58	12.63	12.75	15.02	14.79	19.62	18.83	14.27	13.37	52
7	15.09	13.21	9.94	9.70	12.64	11.82	16.52	16.64	11.85	11.39	33
8	18.74	16.06	15.49	14.77	16.74	16.23	20.51	19.83	14.26	13.45	47
9	18.26	14.71	13.13	13.13	15.52	15.90	19.84	19.03	15.13	14.81	31
10	17.52	14.65	12.53	13.58	14.20	14.86	18.52	17.86	14.47	13.38	66
11	14.85	13.69	11.54	11.08	14.62	13.00	17.77	16.31	14.15	12.85	13
12	14.36	12.36	10.68	10.74	12.46	12.12	15.36	15.36	12.74	11.94	50
13	16.18	12.23	12.32	10.77	11.73	11.59	16.45	15.00	12.68	12.18	22
14	15.90	12.76	12.33	13.38	14.33	13.81	16.76	17.29	13.57	13.29	21
15	19.12	14.67	15.27	14.18	15.09	15.45	19.64	18.48	14.24	14.03	33
16	17.83	14.63	12.50	12.71	16.08	15.33	18.38	17.67	13.67	14.13	24
17	17.27	16.19	14.35	13.96	15.23	15.88	18.00	16.46	12.81	13.27	26
18	18.07	15.11	12.39	12.25	14.60	14.16	18.08	16.74	14.28	13.75	85
Total	17.53	14.77	12.97	12.87	14.74	14.69	18.51	17.91	13.85	13.40	675
Std. Dev.	7.51	6.59	6.46	5.92	5.62	5.82	5.60	6.13	4.38	4.37	

Table 6.2 (continued)
Male Dermatoglyphics—Total Finger Ridge Counts

Group	T		2		3		4		5		N
	R	L	R	L	R	L	R	L	R	L	
1	17.91	13.50	14.27	11.95	15.95	15.59	24.18	22.86	15.64	13.18	22
2	26.04	23.25	20.75	20.38	18.46	19.96	28.54	25.63	15.38	16.50	24
3	23.98	20.53	17.32	18.56	22.51	23.66	32.10	29.59	15.75	14.56	59
4	22.23	21.18	16.95	14.59	15.95	16.23	23.82	22.36	13.91	13.64	22
5	30.78	26.47	23.33	22.91	23.40	25.20	34.11	31.31	19.36	17.58	45
6	26.62	21.10	17.04	17.21	18.54	18.96	29.65	27.85	16.81	14.62	52
7	17.61	14.18	11.61	10.21	14.15	12.18	23.76	19.52	13.00	11.82	33
8	25.15	21.28	20.11	19.81	20.55	21.13	31.26	28.45	16.06	14.74	47
9	25.87	19.68	18.94	17.90	20.61	21.77	29.55	26.94	17.77	15.77	31
10	22.94	18.45	17.09	18.85	17.52	18.32	26.47	22.67	15.58	14.11	66
11	16.85	16.46	14.31	13.69	16.54	15.77	24.23	20.00	14.69	13.62	13
12	18.36	15.38	13.66	12.18	14.54	13.94	21.72	10.02	14.04	12.56	50
13	20.36	14.09	14.23	13.82	13.64	13.14	22.09	18.41	13.73	12.18	22
14	20.76	15.86	16.71	18.57	19.38	18.76	27.10	24.38	17.29	15.33	21
15	27.15	19.58	20.76	21.52	18.06	20.30	30.67	26.88	17.64	16.33	33
16	25.54	21.54	18.54	17.50	21.54	21.00	19.04	15.42	17.63	16.58	24
17	14.31	22.27	19.27	19.00	21.88	22.04	26.73	22.96	15.12	14.12	26
18	24.59	20.31	16.45	16.33	18.61	18.16	27.35	23.04	17.04	15.44	85
Total	23.74	19.57	17.40	17.27	18.71	19.04	27.87	14.82	16.09	14.68	675
Std. Dev.	13.60	11.99	11.08	10.79	11.37	11.65	12.52	12.73	7.31	6.38	

Table 6.2 (continued)
Female Dermatoglyphics—Mainline Patterns

Group	A		B		C		D		N
	R	L	R	L	R	L	R	L	
1	4.95	4.00	4.11	3.05	3.05	2.21	4.84	4.00	19
2	4.65	3.57	4.00	3.04	3.22	1.48	4.74	3.70	23
3	4.73	3.75	4.11	3.48	3.35	2.19	4.63	3.94	79
4	4.68	4.09	4.09	3.50	3.73	2.36	4.73	4.27	22
5	4.55	3.55	3.76	3.02	3.43	2.55	4.43	3.62	42
6	4.67	3.60	4.30	3.42	4.05	2.70	4.97	4.30	40
7	4.82	3.61	3.97	3.45	3.00	2.27	4.52	4.21	33
8	4.44	3.40	3.84	2.93	3.31	2.33	4.64	3.73	45
9	4.62	3.64	4.38	3.54	4.31	2.62	5.23	4.33	39
10	4.28	3.17	3.41	2.96	2.63	1.69	3.99	3.28	78
11	4.65	3.74	3.91	3.78	2.83	2.65	4.43	4.61	23
12	4.51	3.27	3.85	3.44	3.02	2.20	4.44	4.07	41
13	4.83	3.67	4.25	3.33	3.17	2.29	4.92	3.79	24
14	4.52	3.09	4.74	3.61	4.17	2.87	5.17	4.30	23
15	4.95	3.51	4.43	3.57	4.43	2.57	5.16	4.27	37
17	4.86	3.03	4.06	3.44	3.69	2.42	5.03	4.33	36
18	4.73	3.08	3.81	3.19	3.37	2.31	4.49	3.82	93
Total	4.65	3.45	3.99	3.30	3.41	2.29	4.65	3.95	697
Std. Dev.	0.84	1.31	1.22	1.11	1.80	1.61	1.37	1.47	

Table 6.2 (continued)
Female Dermatoglyphics—Maximum Finger Ridge Counts

Group	T		2		3		4		5		N
	R	L	R	L	R	L	R	L	R	L	
1	11.63	11.05	11.95	12.47	13.16	14.32	17.79	17.42	13.42	14.05	19
2	17.26	15.13	15.30	15.00	15.48	16.52	17.96	17.13	13.96	15.74	23
3	14.57	12.46	13.94	13.56	15.68	15.81	19.09	19.76	12.34	11.43	79
4	13.91	13.05	15.32	11.77	15.23	14.14	20.32	20.41	15.00	14.32	22
5	16.74	14.26	15.17	14.45	16.00	16.36	20.38	20.05	13.93	13.26	42
6	17.52	14.47	12.92	12.00	15.70	14.77	19.60	17.42	13.88	13.32	40
7	14.76	11.52	12.30	11.94	14.03	12.58	17.76	14.79	11.18	10.61	33
8	15.47	13.78	13.80	14.16	14.58	14.29	18.84	18.96	13.78	13.42	45
9	14.90	13.33	13.46	12.56	15.36	15.64	19.74	19.15	14.15	13.49	39
10	16.51	13.82	14.00	13.40	14.24	13.90	19.01	18.63	14.24	13.03	78
11	13.22	11.09	12.00	11.22	12.78	13.13	16.43	16.04	12.52	11.61	23
12	14.15	13.15	11.78	11.15	14.37	12.78	17.61	16.29	13.32	12.29	41
13	13.29	10.33	11.96	10.67	12.75	13.38	16.67	16.71	11.79	11.04	24
14	16.74	13.17	15.00	13.65	16.04	15.70	18.39	18.43	14.74	13.30	23
15	14.84	12.95	11.89	12.38	14.05	13.19	16.76	16.08	12.89	12.49	37
17	14.72	13.47	12.58	12.08	14.89	13.19	17.94	17.78	13.56	12.81	36
18	16.90	13.82	12.97	12.43	14.41	13.51	18.34	17.22	14.26	13.10	93
Total	15.45	13.20	13.35	12.77	14.72	14.30	18.55	17.95	13.52	12.79	697
Std. Dev.	7.34	6.69	6.73	6.02	5.65	6.21	5.76	6.45	4.35	4.54	

Table 6.2 (continued)
Female Dermatoglyphics—Total Finger Ridge Counts

Group	T R	T L	2 R	2 L	3 R	3 L	4 R	4 L	5 R	5 L	N
1	15.11	15.42	17.05	16.63	17.21	19.58	26.37	23.68	15.00	14.26	19
2	24.22	19.09	22.04	22.61	22.39	22.48	29.09	27.00	15.87	16.39	23
3	18.66	15.99	18.48	18.54	20.35	21.34	27.67	28.16	13.35	12.27	79
4	18.50	18.64	21.05	16.73	19.77	19.09	28.64	27.50	16.18	15.91	22
5	21.90	20.52	21.74	22.07	22.17	14.12	30.48	29.38	16.26	15.43	42
6	23.00	19.15	16.35	16.88	20.63	18.65	28.97	25.85	15.77	14.65	40
7	18.45	14.70	16.36	15.58	16.88	14.70	23.64	19.42	13.79	12.61	33
8	20.18	18.49	18.36	20.02	18.04	17.73	27.84	26.29	16.51	15.29	45
9	20.92	18.69	17.72	16.87	20.49	21.79	28.28	17.77	15.54	14.90	39
10	23.72	20.12	19.45	18.87	18.50	18.17	27.60	24.42	14.63	13.53	78
11	17.39	13.61	14.61	13.17	14.91	18.39	22.43	21.17	13.52	12.61	23
12	18.17	17.49	15.15	13.71	16.68	15.71	23.54	22.90	14.34	13.27	41
13	15.58	12.75	15.33	13.88	14.17	17.04	22.21	21.17	13.17	11.21	24
14	24.87	20.43	21.09	17.26	20.87	21.70	27.96	26.83	17.22	15.17	23
15	19.41	19.41	15.73	16.59	15.70	16.65	22.62	20.65	13.84	13.30	37
17	23.06	20.14	18.31	16.89	19.83	17.81	27.75	25.11	24.53	13.69	36
18	21.06	17.86	17.68	16.39	18.20	18.11	26.45	24.18	16.40	15.13	93
Total	20.58	18.02	18.06	17.44	18.78	18.97	26.78	15.03	15.05	14.04	697
Std. Dev.	12.88	12.00	11.84	11.16	11.50	12.11	12.59	12.84	6.19	6.24	

tion of the right palm than the left. As for sex difference, the same level and same pattern of intercorrelations seem to hold. There may be a tendency for the D mainline in females to be slightly more correlated with other variables than in males, but this is tenuous.

As for the finger ridge counts, the maximum and total counts for each finger are so highly correlated that it would seem to make little difference which

Table 6.3
Correlation Matrix for Male and Female Dermatoglyphic Variables
(Males in lower left triangle, females in upper right)

	1	2	3	4	5	6	7	8	9	10
1	1.00	.48	.50	.40	.30	.11	.43	.37	-.04	-.05
2	.51	1.00	.30	.37	.15	.06	.25	.34	-.09	-.09
3	.44	.32	1.00	.59	.66	.22	.85	.60	-.01	-.05
4	.43	.45	.53	1.00	.39	.30	.57	.87	-.02	-.04
5	.32	.24	.64	.41	1.00	.27	.63	.41	.02	.01
6	.17	.16	.24	.33	.26	1.00	.23	.41	.04	.03
7	.44	.30	.84	.50	.65	.25	1.00	.62	.00	-.03
8	.42	.40	.53	.83	.44	.45	.57	1.00	-.01	-.03
9	-.07	-.05	-.05	-.04	-.03	.00	-.05	-.05	1.00	.76
10	-.06	-.05	-.04	-.05	-.02	-.05	-.04	-.05	.76	1.00
11	.02	-.01	.04	-.02	.00	-.04	.03	-.01	.46	.50
12	.01	.02	.01	.03	-.02	-.02	.00	-.02	.49	.51
13	.04	.02	.02	.00	.02	-.07	.01	-.04	.43	.46
14	.05	.03	.04	.06	.03	-.07	.04	.02	.40	.45
15	.01	.04	.01	.01	.03	-.07	.01	-.02	.41	.40
16	.01	.05	.04	-.02	.01	-.08	.01	-.04	.41	.42
17	.01	.00	.00	-.07	.00	-.08	-.03	-.10	.40	.40
18	.01	-.01	-.01	-.04	-.01	-.07	-.04	-.08	.43	.46
19	-.05	-.05	-.03	-.05	-.01	-.01	-.04	-.06	.87	.73
20	-.08	-.04	-.05	-.08	-.04	-.05	-.06	-.07	.67	.88
21	-.01	.01	.01	-.02	.00	-.04	.01	-.02	.44	.49
22	.01	.03	.01	.00	-.01	-.05	.00	-.02	.41	.47
23	.05	.07	.01	-.01	.03	-.08	-.01	-.05	.36	.41
24	.05	.06	.05	.06	.05	-.03	.04	.04	.36	.41
25	.03	.06	.06	.05	.03	-.07	.03	.02	.40	.41
26	.00	.00	.02	-.02	-.01	-.08	-.02	-.05	.39	.42
27	.05	.03	.04	-.03	.07	-.05	.02	-.04	.37	.40
28	.01	-.01	-.02	-.04	-.01	-.08	-.03	-.07	.39	.42

variable was used. Previously, Parsons (1964) reported correlations on the order of .80 for pattern intensity indices and ridge counts among Australian aboriginals, but the present survey correlations are of that order or higher. For both sexes the correlations of each finger count are highest with the corresponding finger on the opposite hand without exception (unlike the pattern on the palm, the correspondence is closer with bilateral homologues or antemeres than

Table 6.3 (continued)

11	12	13	14	15	16	17	18	19	20
.06	.03	.07	.08	.10	.12	.00	-.03	-.01	-.01
.06	.06	.10	.06	.10	.09	.04	.03	-.07	-.07
.02	-.01	.05	.02	.01	.01	-.01	-.04	-.01	-.03
.00	-.01	.04	.03	.01	.02	-.04	-.05	-.02	-.03
.05	.04	.10	.07	.07	.06	.04	.02	.02	.02
.04	.01	.05	.06	.06	.02	.02	.05	.04	.05
.05	.02	.08	.04	.02	.02	-.02	-.03	-.02	-.03
.03	.00	.06	.04	.01	.01	-.03	-.01	-.04	-.02
.49	.48	.46	.45	.38	.33	.41	.42	.88	.70
.49	.50	.46	.48	.38	.35	.42	.44	.74	.90
1.00	.71	.68	.63	.52	.55	.43	.46	.49	.47
.71	1.00	.62	.64	.48	.53	.44	.47	.47	.47
.70	.63	1.00	.80	.62	.62	.47	.46	.46	.45
.66	.67	.78	1.00	.61	.65	.47	.49	.44	.47
.54	.55	.64	.64	1.00	.73	.58	.53	.43	.41
.57	.58	.64	.71	.76	1.00	.52	.54	.39	.38
.44	.43	.47	.50	.56	.56	1.00	.76	.45	.45
.51	.49	.53	.56	.58	.60	.75	1.00	.44	.45
.47	.49	.45	.39	.44	.44	.43	.46	1.00	.79
.46	.48	.43	.40	.40	.42	.41	.45	.78	1.00
.89	.69	.66	.65	.54	.56	.44	.50	.49	.49
.71	.88	.62	.64	.54	.56	.42	.48	.46	.48
.66	.60	.88	.73	.64	.64	.47	.51	.42	.41
.62	.63	.73	.87	.64	.68	.50	.54	.41	.41
.60	.61	.68	.70	.87	.77	.56	.59	.46	.43
.59	.60	.65	.70	.69	.87	.54	.58	.46	.45
.47	.44	.49	.52	.52	.55	.84	.69	.46	.44
.49	.46	.49	.51	.50	.54	.66	.87	.44	.49

with neighboring fingers). Besides this, correlations are highest with those neighboring fingers on the same hand, and on the neighboring homologues on the opposite hand. Females seem to have slightly higher intercorrelations over-all. One apparent difference is that males have one area of higher intercor-relations, in the third and fourth maximum counts on the left hand. Otherwise, the thumbs show the lowest correlations, while the middle and fourth fingers show the highest.

Table 6.3 (continued)

	21	22	23	24	25	26	27	28
1	.06	.04	.08	.10	.12	.11	.02	.00
2	.06	.08	.08	.07	.10	.08	.03	.05
3	.01	−.04	.06	.02	.02	.01	−.02	−.04
4	.00	−.03	.03	.00	.01	.00	−.07	−.04
5	.05	.00	.10	.05	.07	.06	.03	−.01
6	.02	−.02	.06	.04	.05	.02	−.04	.03
7	.04	−.02	.09	.02	.03	.01	−.01	−.02
8	.02	−.01	.06	.02	.02	.00	−.06	−.01
9	.47	.46	.44	.44	.38	.35	.42	.39
10	.48	.48	.45	.48	.39	.38	.43	.41
11	.90	.69	.66	.65	.57	.57	.47	.42
12	.70	.89	.59	.63	.55	.56	.49	.44
13	.67	.61	.88	.76	.68	.66	.49	.42
14	.63	.62	.73	.88	.66	.69	.47	.43
15	.53	.49	.60	.61	.87	.67	.50	.45
16	.58	.55	.62	.63	.73	.86	.50	.47
17	.44	.44	.46	.48	.58	.51	.85	.68
18	.47	.47	.46	.47	.55	.55	.69	.87
19	.51	.48	.49	.47	.45	.39	.47	.41
20	.49	.49	.48	.51	.44	.42	.47	.43
21	1.00	.75	.72	.69	.61	.60	.48	.42
22	.75	1.00	.64	.68	.58	.60	.49	.45
23	.69	.67	1.00	.80	.72	.68	.51	.42
24	.65	.68	.79	1.00	.70	.74	.51	.43
25	.62	.61	.72	.73	1.00	.78	.57	.49
26	.62	.62	.69	.75	.80	1.00	.56	.52
27	.50	.47	.52	.56	.59	.61	1.00	.72
28	.51	.48	.49	.52	.56	.59	.73	1.00

To recapitulate, the ridge patterns on the fingers show stronger bilateral correspondences than they do correspondences among neighboring fingers, while the palm patterns show tendencies toward some correspondences among lines from the same hands, although the right hand shows closer correspondences than the left. Among the fingers, the thumb shows the least correlation with the fourth finger, which is otherwise most highly correlated with the patterns of other fingers. The thumb seems to be the focus of one central pattern stage, and the fourth finger the focus of another.

As for the differences among the different groups, these are best described by a one-way analysis of variance, comparing the variation within each village with the variation among villages. From Table 6.4, there are a few notable points to make. First, the F-ratios for males are generally higher than for females, meaning that males show a higher degree of variation with respect to the separate variables than do females in the sample. The highest female scores for diversity are mainline scores for the right hand. The order runs: mainlines C, B, and D of the right hand, then mainline D of the left; then the maximum ridge counts of the fifth and fourth fingers of the left hand, and then the left and right mainline A scores. In the males, it is true that the right mainlines are more variable (B the most, then D, and C rather far down the list), but ridge counts for the second, third, and fourth fingers of the left hand, as well as the fourth finger of the right, are also quite variable from one village to the next.

It is interesting that those patterns noted in the correlation matrices (the high intercorrelations of the right mainlines, especially in females, the high correlations of the left third and fourth fingers in males) reappear in the analysis of variance results. That is, those variables which are most correlated with others from the same individual are also the most variable among the different populations. There is the possibility that this is a reflection of relative canalization of the various traits, with the most genetic control centered in the area of the left second and third, and right fourth fingers, as well as the right mainlines B, C, and D, which form patterns at the bases of the lateral three fingers for males; in females, the areas of interest are in the same palmar region as well as the left fourth and fifth fingers. There is the implication, then, that the differences among groups are genetically based ones. The fact that females seem less variable from one village to the next is also interesting from a genetic point of view, in that such differences have in the past been attributed to greater developmental homeostasis in females. Whatever the reason, it is a common finding in many other traits that males show more variation than females. But here it is interesting especially because the dermatoglyphic traits are determined so early in life and are not influenced by adolescent growth differentials.

Table 6.4
F-Ratios for Male Dermatoglyphic Variables

Variables			F-ratio	Probability
Mainlines	A	1	1.70	$.38 \times 10^{-1}$
		2	2.60	$.65 \times 10^{-3}$
	B	1	3.99	$.48 \times 10^{-5}$
		2	1.49	$.92 \times 10^{-1}$
	C	1	3.03	$.12 \times 10^{-3}$
		2	1.61	$.56 \times 10^{-1}$
	D	1	3.51	$.21 \times 10^{-4}$
		2	1.74	$.32 \times 10^{-1}$
Max. count	T	R	2.09	$.66 \times 10^{-2}$
		L	2.57	$.75 \times 10^{-3}$
	2	R	2.73	$.38 \times 10^{-3}$
		L	3.36	$.34 \times 10^{-4}$
	3	R	2.82	$.26 \times 10^{-3}$
		L	3.36	$.35 \times 10^{-4}$
	4	R	3.36	$.34 \times 10^{-4}$
		L	2.69	$.45 \times 10^{-3}$
	5	R	1.59	$.61 \times 10^{-1}$
		L	2.11	$.57 \times 10^{-2}$
Total count	T	R	2.59	$.68 \times 10^{-3}$
		L	2.85	$.23 \times 10^{-3}$
	2	R	2.51	$.95 \times 10^{-3}$
		L	3.62	$.14 \times 10^{-4}$
	3	R	2.37	$.18 \times 10^{-2}$
		L	3.51	$.21 \times 10^{-4}$
	4	R	3.04	$.11 \times 10^{-3}$
		L	3.20	$.61 \times 10^{-4}$
	5	R	1.85	$.20 \times 10^{-1}$
		L	2.14	$.52 \times 10^{-2}$

Discriminant Analysis

These twenty-eight variables from the 17 villages (and one conglomerate male group) were subjected to a discriminant analysis in similar fashion to the anthropometric measurements. These sets of data needed no estimation of missing variables, or, rather, it was not attempted, because these are not parametric variables with normal distributions. If an individual had missing fingers

Table 6.4 (continued)

Variables			F-ratio	Probability
Mainlines	A	1	2.10	$.73 \times 10^{-2}$
		2	2.16	$.55 \times 10^{-2}$
	B	1	2.83	$.33 \times 10^{-3}$
		2	2.06	$.87 \times 10^{-2}$
	C	1	3.43	$.38 \times 10^{-4}$
		2	1.75	$.34 \times 10^{-1}$
	D	1	2.78	$.40 \times 10^{-3}$
		2	2.40	$.19 \times 10^{-2}$
Max. count	T	R	1.61	$.60 \times 10^{-1}$
		L	1.13	.32
	2	R	1.14	.31
		L	1.37	.15
	3	R	1.07	.38
		L	1.62	$.58 \times 10^{-1}$
	4	R	1.42	.12
		L	2.24	$.39 \times 10^{-2}$
	5	R	2.01	$.11 \times 10^{-1}$
		L	2.37	$.22 \times 10^{-2}$
Total count	T	R	1.61	$.60 \times 10^{-1}$
		L	1.32	.18
	2	R	1.23	.24
		L	1.84	$.23 \times 10^{-1}$
	3	R	1.42	.13
		L	1.67	$.48 \times 10^{-1}$
	4	R	1.43	.12
		L	1.90	$.17 \times 10^{-1}$
	5	R	1.67	$.47 \times 10^{-1}$
		L	1.87	$.20 \times 10^{-1}$

or a missing hand, he was excluded from the analysis. The remaining samples are still the largest in the survey. Judging by Tables 6.5a and 6.5b the discriminations for the male and female sets are both very poor.

It is important to realize that the H_1 tests of the equality of dispersions for the discriminant analyses of both sexes give highly significant results, so that it is questionable whether discriminant analysis, or any multivariate technique assuming normality of dispersions, is appropriate. Rao's F test of significance suggests that the probability of achieving such discrimination by chance is $.561 \times 10^{-5}$ for males and $.402 \times 10^{-6}$ for females. This is surprising in that the univariate results showed a number of male variables to be more "discriminatory" than their female counterparts. Figures 6.2 and 6.3 present the plots of the group centroids in discriminant space for the two sexes for the first four functions in females and first three for males given in Tables 6.6a and 6.6b.

For the females, the first discriminant places villages 17 and 18 (the Austronesian speakers) at one extreme, and village 3 at the other. Looking at the discriminant loadings (Tables 6.7a and 6.7b) to discover the best interpretation of the nature of this discriminant, it appears that the first function is most closely associated with ridge counts, except for the first and fifth fingers, and emphasizing the third and fourth fingers of the left hand. These are contrasted with the mainlines (given negative correlations), especially C and D, and also the right A mainline as well. The left A mainline has a positive correlation, while the right A is negative. This pattern reflects quite satisfactorily the preponderance of high F's among mainlines and the left fourth fingers in females.

The second discriminant, almost as important as the first, serves to separate groups 1 and 2 from the clustered center, along with 17 again and village 10. The Eivo, except for village 4, score at the lower extreme on this discriminant. The highest loadings on this axis are with the maximum count for the left fifth finger, and secondarily with the left mainlines C, B, and D, all in a negative way. While the first discriminant emphasizes the third and fourth fingers of the left hand, the second emphasizes the fifth. In the mainlines, the shift is from the A mainline to the B mainline termination, near the fifth finger, with lower mainline scores going with higher maximum ridge counts (greater complexity

Table 6.5a
Relative Importance of Male Dermatoglyphic Discriminant Functions

Wilks's Lambda for total discrimination = .378
Rao's F = 1.48
DF_1 = 448 DF_2 = 9155.7
Chance probability of F = .40 x 10^{-6}

Decreasing probability of significance of successive roots

Root	Eigenvalue	Percent of trace	Chi-square tests of significance with successive roots removed	
			χ^2	Probability
0			655.4	1 x 10^{-9}
1	.1651	16.1	552.5	.137 x 10^{-5}
2	.1390	13.6	464.8	.268 x 10^{-3}
3	.1315	12.9	381.6	.167 x 10^{-1}
4	.1118	10.9	310.2	.175
5	.0959	9.4	248.0	.567

Test of H_1 (equality of dispersions):
  F = 1.911; associated probability = .173 x 10^{-18}
  DF_1 = 6045
  DF_2 = 104628

Table 6.5b
Relative Importance of Female Dermatoglyphic Discriminant Functions

Wilks's Lambda for total discrimination = .370
Rao's F = 1.37
DF_1 = 476 DF_2 = 9245.0
Chance probability of F = .56 x 10^{-5}

Decreasing probability of significance of successive roots

Root	Eigenvalue	Percent of trace	Chi-square tests of significance with successive roots removed	
			χ^2	Probability
0			647.4	.298 x 10^{-6}
1	.164	15.4	548.5	.118 x 10^{-3}
2	.159	15.3	452.6	.155 x 10^{-1}
3	.131	12.6	372.4	.200
4	.106	10.1	307.0	.569

Test of H_1 (equality of dispersions)
  F = 1.733; associated probability = .260 x 10^{-17}
  DF_1 = 6045
  DF_2 = 131271

6.2 Male dermatoglyphics: centroids of the 18 samples on the first three discriminant axes

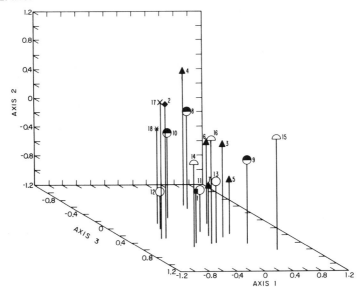

6.3 Female dermatoglyphics: centroids of the 17 samples on the first three discriminant axes

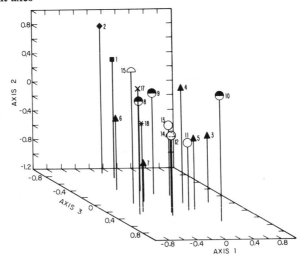

Table 6.6a
Dermatoglyphics: Group Centroids in Male Discriminant Space

Village	Function 1	Function 2	Function 3	Function 4
1.	-0.27057	-0.48401	0.41056	-0.37930
2.	-0.62799	0.65541	0.26773	0.75587
3.	0.21367	0.09906	-0.25397	-0.30485
4.	0.40598	-0.66415	-0.63021	-0.11482
5.	0.39873	0.45352	0.16027	0.02616
6.	0.29961	-0.06920	-0.10479	0.02966
7.	0.04466	-0.73518	0.23810	0.07884
8.	0.40199	0.14055	-0.56914	0.18105
9.	0.40624	0.04387	0.42398	-0.20862
10.	-0.09203	-0.02326	-0.24356	0.36961
11.	0.02293	-0.63019	0.12117	0.09723
12.	-0.45549	-0.47555	0.01739	0.08277
13.	-0.02103	-0.35368	0.41095	0.49028
14.	-0.42125	-0.05413	0.55608	0.01706
15.	0.65938	0.34693	0.62614	0.21688
16.	-0.05298	0.22517	0.42677	-0.83512
17.	-0.67691	0.68747	-0.29056	-0.19902
18.	-0.37673	0.12784	-0.15367	-0.22975

Table 6.6b
Dermatoglyphics: Group Centroids in Female Discriminant Space

Village	Function 1	Function 2	Function 3	Function 4
1.	-0.03529	0.43036	-0.75191	0.06364
2.	-0.12545	0.84725	-0.91274	-0.48784
3.	0.62238	-0.26293	-0.00384	0.05535
4.	0.26548	0.04011	-0.00383	0.30274
5.	0.28155	-0.21558	-0.19518	-0.08588
6.	-0.30109	-0.18361	-0.39778	-0.03475
7.	-0.28429	-0.67304	-0.04089	-0.48351
8.	0.00127	0.05440	-0.37712	-0.17811
9.	0.11447	0.02125	-0.25342	0.55497
10.	0.38851	0.53656	0.48276	-0.27930
11.	0.19138	-0.26232	0.19993	0.12453
12.	-0.16266	-0.08887	0.34297	0.01273
13.	0.31879	-0.23691	-0.20947	0.07782
14.	-0.07920	-0.14291	0.25804	0.70575
15.	-0.40104	0.06358	-0.01005	0.54502
17.	-0.61167	0.55778	0.27251	0.34720
18.	-0.46647	-0.19438	0.22921	-0.28025

in the fingers is then associated with greater longitudinality in the mainlines). There is also a heavier correlation of left scores than right *in general.*

The third discriminant most neatly divides the northern villages from the south, with villages 1 and 2 again at one extreme, and only villages 13 and 15 scoring negatively among the southern villages. This is more closely associated with mainlines than with finger-ridge counts. After this, the patterns become less distinct.

Table 6.7a
Discriminant Loadings for Male Dermatoglyphic Variables

| Variables | Discriminant functions | | | |
	1	2	3	4
1	.21	−.22	.18	−.33
2	.32	−.38	.19	−.12
3	.41	−.21	.65	.00
4	.18	−.39	.15	.03
5	.45	−.12	.41	−.17
6	.37	−.19	.00	.25
7	.38	−.16	.62	.08
8	.22	−.33	.16	.12
9	.25	.47	−.07	.10
10	.21	.48	−.28	.07
11	.28	.57	−.05	.18
12	.23	.66	−.12	.19
13	.36	.45	−.21	−.28
14	.32	.61	−.18	−.12
15	.50	.49	−.18	−.05
16	.45	.37	−.15	.01
17	.26	.26	−.08	−.05
18	.14	.52	.09	−.03
19	.30	.55	−.03	−.04
20	.21	.55	−.21	−.09
21	.27	.56	−.01	.06
22	.29	.68	−.05	.14
23	.19	.50	−.11	−.30
24	.30	.62	−.11	−.23
25	.41	.52	−.07	−.14
26	.44	.46	−.08	−.16
27	.19	.38	.15	−.19
28	.16	.49	.13	−.12

The male discriminant functions are similarly indistinct and difficult to interpret. The first discriminant function places villages 2 and 17 at one extreme and 15 at the other. The first discriminant is quite different from the female counterpart in its associations with the single variables, although it is quite easily interpreted. Except for mainline A, every right maximum ridge count and mainline score is higher than its left antemere, but this is especially true for the mainlines B, C, and D. There are no negative correlations at all, quite a dis-

Table 6.7b
Discriminant Loadings for Female Dermatoglyphic Variables

Variables	Discriminant functions			
	1	2	3	4
1	-.24	-.21	-.20	.20
2	.28	-.17	-.39	.15
3	-.12	-.21	-.31	.59
4	-.05	-.28	.05	.44
5	-.25	-.13	-.25	.63
6	-.17	-.33	-.09	.39
7	-.23	-.09	-.36	.58
8	-.23	-.25	-.13	.46
9	-.12	.06	.06	-.23
10	-.10	.18	-.01	-.11
11	.21	.10	-.07	-.05
12	.17	.15	-.14	-.12
13	.08	-.03	-.09	.09
14	.27	.03	-.27	.09
15	.19	.00	-.07	-.03
16	.38	.11	-.04	.13
17	-.10	.29	.08	.04
18	-.14	.41	-.25	-.01
19	-.09	.27	.11	-.06
20	-.09	.27	.08	.07
21	.15	.18	-.05	-.06
22	.18	.22	-.21	-.16
23	.10	.09	-.15	.02
24	.25	.04	-.21	.13
25	.12	.15	-.13	-.04
26	.25	.07	-.16	.09
27	-.20	.05	-.10	-.06
28	-.20	.15	-.19	-.05

tinction from the female set. In this sense the first discriminant is a "size" vector. There also seems to be some accentuation of differences in the hands with respect to loops and whorls. Individuals with loops on the right hand and whorls on the left will score highest, while the reverse will be lowest (the left total ridge counts have a higher correlation for the middle three fingers).

The second discriminant function separates out villages 2 and 17 once again at one extreme, contrasted to villages 4, 7, and 11. Again, these are not familiar combinations. The loadings contrast ridge counts (but especially those from the left second and third fingers) with the mainlines (but especially A, B, and D). Except for the fourth finger, the *left* hand determines the scores on this discriminant. In a real sense, then, the first male discriminant is dealing with the right hand, and the second is dealing with the left. The patterning is highly interesting, but the results for village discrimination are once again disappointing.

The third discriminant is closely tied to the B, C, and D mainlines of the right hand and nothing else. It separates out groups 14 and 15, the Siwai, from 4 and 8 at the other extreme.

Therefore, the different discriminants for the males and females have some vague resemblance to one another, but not in a very clear fashion. The first discriminant is different in the two sexes, being right-handed in males (especially the mainlines and third and fourth fingers), and simply stressing the left third and fourth fingers as well as the last three mainlines in females. Both discriminants place group 17 at one extreme, but that is as far as it goes.

The second discriminant in both sexes contrasts mainline scores with finger-ridge patterns, with the left hand predominating in both sexes. In females, the most important finger is the fifth, while in males it is the second, third, and fifth. In both sexes, the second discriminant separates out village 7 at one extreme and villages 2 and 17 at the other.

The third discriminant is mainly associated with the right B and D mainlines, and little else.

The percentage of correct assignments of individuals to the different groups according to the centroid rule is strikingly similar in both sexes, being .246 for the females (with 17 groups) and .245 for the males (with 18 groups). This is, of course, lower than the discriminatory power of the anthropometrics. The small dispersion of the village means reflects the poor discrimination.

There are patterns of heritable dermatoglyphic variation which differ among these groups. The variations, or the most controlled portion of the dermato-glyphics by the genes, seem to be along the middle of the palm and fingers—that is, centering on the third and fourth fingers, and the B, C, and D mainlines, which usually tend to terminate near the bases of these fingers. How susceptible to generalization these results are is a highly provocative question. I think this may be extremely important for dermatoglyphic students to explore.

Perhaps the dermatoglyphics show such weak discriminatory power not because they are weakly controlled by genetic variation, but because they are compromises of genetic influences from a very large number of loci. In other words, the magnitude of diversification which characterizes these Bougainville villages over the past 2,000 years or more has not been enough to change the overall dermatoglyphic patterns appreciably. More strictly speaking, the normal variation in dermatoglyphics is so great within such groups that the genetic differences affecting these traits which have arisen in Bougainville are comparatively unimportant.

7 Dental Variation

Because the teeth are known to vary in size and shape from one major population to the next, and because dental variation is generally assumed to be relatively closely tied to genotypic variation, I was anxious to obtain a series of dental measurements from these villagers once it became clear how great the variation in other characters was. Fortunately, Dr. Howard Bailit of the University of Connecticut was more than willing to visit Bougainville in 1970 to take dental impressions, and to collaborate in this aspect of the study.

Very few dental surveys have included comparisons with other sorts of biological variation. Littlewood (1973), in a New Guinea survey, found general correspondence in his comparison of anthropometric, blood group, and dental data. Similarly, Sofaer et al. (1972) reported a moderately good correspondence between their dental observations (e.g., shoveling of incisors, molar cusp number, and fissure pattern) and genetic distance as determined by blood group gene frequencies in Papago Indians of the southwest.

Because metric data are amenable to more powerful multivariate statistical treatment, and because qualitative observations on dental morphology showed comparatively little variation in the Bougainville series, it suffices here to consider only measurements on the length and breadth of the individual teeth.

The teeth are in the process of development over an extended but discrete period of time. The first radiographic evidence of the dentition is usually radiographically detectable by the 18th week of gestation, and the third molars do not erupt until the age of eighteen or nineteen in most Caucasian populations (Kraus and Jordan 1965). However, the size or shape of teeth is subject to environmental influences primarily up to the point of their calcification. Once the crown of a tooth has calcified, environmental factors such as systemic disease or nutrition are not going to affect it. Of course, after eruption local factors such as caries, trauma, or tooth wear will influence tooth size.

A study by Moorrees, Fanning, and Hunt (1963) showed that, excluding third molars, all teeth have either completed or started crown calcification by age eight or nine. This means that it is only during the prenatal period and the first seven years of life that the size of the permanent teeth will be affected by the environment in their development. In fact, several studies in mice and humans suggest that the prenatal period is most critical and, more specifically, that maternal physiology is a major determinant of tooth size (Main 1966, Bailit and Sung 1968, Tenczar and Bader 1966).

Two approaches have been used to determine the relative genetic and environmental components of tooth size. Using measurements on fraternal and identical twins, Osborne et al. (1958) found that genetic variance accounted for over half of the total phenotypic variance in the mesio-distal (M-D) diameters of the permanent anterior teeth.

Working with dental casts of Pima Indians, Potter et al. (1968) performed a factor analysis on 56 tooth diameters, and then looked at the covariances of three factors among relatives. They reported that there are significant environmental effects on tooth size as well.

Neither result is unexpected in a continuous character such as tooth size, which has a discrete interval for development and thereafter is not susceptible to marked environmental changes.

In related papers (Friedlaender 1974, Dietz et al. 1974), the dental measurements for the Bougainville series have been used to estimate the heritability of these characters in this population. These studies indicate in a general way that most measurements are approximately as heritable as finger ridge counts. Comparing the pattern of variation in the dentition over these villages, then, should provide for an interesting comparison with the previously analyzed sets of data.

In terms of variation in size, the tooth diameters most commonly examined are the mesio-distal (M-D) and labio-lingual (L-L). These are obvious, easy, and reliable measurements to take. The M-D diameter is defined as the greatest diameter of the tooth crown measured parallel to the occlusal and labial surfaces. The L-L diameter is defined as the greatest diameter of the tooth measured perpendicular to the M-D diameter. Of course, there are many other measures of tooth size, such as crown height or root length, but it is difficult and often impossible to obtain these dimensions from dental casts.

The description of the measuring techniques and data preparation are given in the Appendix and follow the anthropometric scheme in broad outline. Since the teeth are bilaterally symmetrical, that is, the mean difference between antimeres is not significantly different from zero, it is largely unnecessary to measure both left and right sides. Instead of measuring 32 teeth, 16 are adequate to describe most tooth size variation in a population. Needless to say, this results in substantial savings of time and money. In this survey, the teeth on the left were measured preferentially, and if any were missing, their antimeres on the right were substituted. When the measuring was completed, it was clear that a number of teeth were so frequently missing that it would be unwise to include them in the multivariate analysis. These were, for both mesio-distal and labio-lingual diameters, M^1, M^3, I_2, M_2, and M_3. The labio-lingual diameter on P^1 was also excluded. For the remaining teeth, missing values were estimated from other, usually adjacent, teeth in the same individual.

Univariate and Bivariate Statistics

As mentioned above, Howard Bailit and Glen Rapoport, a dental student at the University of Connecticut, spent the summer of 1970 taking dental impressions of the people I had sampled three summers previously. Of the 18 villages visited by me, 14 were also visited by Bailit and Rapoport. Villages 14, 15, and 16 from the Siwai area and village 11, a small group of Nasioi-speakers, were not included in the dental study. Only those juveniles and adults previously surveyed were included in this survey.

Tables 7.1a and 7.1b present the means of the M-D and L-L diameters of the teeth from each of the 14 villages for males and females. Just by looking at the averages for both sexes, it is evident that there is a north-south gradient in posterior tooth size among the villages, with those in the north of the series generally smaller. However, the canines and incisors show little variation. By the same token, there is considerable variation in tooth size within language groups. Generally, the male means are larger than those of the corresponding female, particularly for the canines and cheek teeth.

The standard deviations among the 14 villages vary little, except for small increases for certain teeth, such as maxillary lateral incisors or second molars. This, of course, is to be expected, and the standard deviations are approximately the same as those reported in other odontometric studies—generally between .4 and .7 (Bailit, and Sung 1968).

The correlation matrix for the tooth diameters are presented in Table 7.2 for males (lower left triangle) and for females (upper right). The highest correlations are, almost without exception, between pairs of either mesio-distal

Table 7.1a
Tooth Size Means for Males

Mesio-Distal

Village	Upper teeth						Lower teeth				
	I^1	I^2	C^1	P^1	P^2	M^2	I_1	C_1	P_1	P_2	M_1
1	8.6	6.9	8.2	7.1	6.6	9.3	5.3	7.0	7.1	7.0	11.3
2	8.5	6.8	8.0	6.7	6.2	9.3	5.2	6.9	6.7	6.9	11.4
3	8.9	7.2	8.2	7.1	6.6	9.6	5.3	7.2	7.1	7.0	11.6
4	8.9	7.2	8.3	7.2	6.5	9.4	5.4	7.1	7.0	6.9	11.4
5	8.7	7.2	8.3	7.2	6.4	9.4	5.3	7.0	7.0	6.8	11.3
6	9.0	7.4	8.2	7.3	6.8	9.1	5.6	7.2	7.1	7.2	11.6
7	9.2	7.7	8.3	7.3	6.8	9.8	5.7	7.2	7.1	7.2	11.5
8	8.8	7.3	8.1	7.0	6.5	9.3	5.6	7.0	6.9	6.8	11.5
9	8.5	7.1	8.0	7.3	6.8	9.4	5.4	6.9	7.2	7.2	11.5
10	8.6	7.3	8.0	7.2	6.7	9.3	5.3	7.0	7.0	7.0	11.6
12	8.8	7.3	8.2	7.5	7.0	9.8	5.4	7.1	7.2	7.6	11.5
13	8.6	7.6	8.4	7.5	6.9	9.9	5.5	7.2	7.4	7.5	11.7
17	8.6	6.9	8.0	7.5	6.8	9.4	5.3	6.8	7.1	7.6	11.6
18	8.8	7.2	8.1	7.6	7.0	9.8	5.4	7.0	7.4	7.4	11.9

Labio-Lingual

Village	Upper teeth					Lower teeth					
	I^1	I^2	C^1	P^2	M^2	I_1	C_1	P_1	P_2	M_1	N
1	7.2	6.5	8.5	9.9	12.0	5.8	7.5	8.0	8.4	10.9	20
2	7.2	6.4	8.4	9.9	11.6	5.9	7.2	7.7	8.2	10.7	43
3	7.8	6.8	8.8	10.2	12.4	6.4	7.6	8.4	8.8	11.3	22
4	7.3	6.5	8.4	9.5	12.1	5.9	7.7	8.1	8.7	11.1	41
5	7.5	6.6	8.8	9.9	11.8	6.1	7.6	8.0	8.7	11.0	34
6	7.8	6.9	9.0	10.2	12.0	6.4	7.8	8.3	8.7	11.0	22
7	7.7	7.1	9.0	10.2	12.3	6.3	7.8	8.3	8.8	11.1	47
8	7.6	6.7	8.5	9.9	11.9	6.1	7.5	8.1	8.4	11.0	29
9	7.6	6.6	8.3	10.2	11.9	6.5	7.7	8.1	8.5	11.0	45
10	7.6	6.7	8.6	10.0	12.2	6.5	8.0	8.1	8.6	11.1	28
12	7.7	6.9	8.9	10.4	12.4	6.3	8.0	8.3	9.0	11.2	14
13	7.6	7.1	8.8	9.9	12.3	6.3	7.9	8.5	8.7	11.0	17
17	7.4	6.5	8.4	10.0	11.6	6.0	7.4	8.1	8.6	10.9	41
18	7.6	6.8	8.8	10.3	12.4	6.4	7.9	8.5	8.8	11.3	418

Table 7.1b
Tooth Size Means for Females

	Mesio-Distal										
	Upper teeth						Lower teeth				
Village	I^1	I^2	C^1	P^1	P^2	M^2	I_1	C_1	P_1	P_2	M_1
1	8.4	6.8	8.1	6.8	6.3	9.7	5.2	6.5	6.8	6.7	10.8
2	8.6	7.3	7.8	7.0	6.4	8.9	5.4	6.8	6.8	7.0	11.1
3	8.7	7.0	7.9	6.8	6.3	9.6	5.2	6.7	6.7	6.7	11.0
4	8.7	6.9	7.9	6.8	6.4	9.3	5.3	6.7	6.9	7.0	11.1
5	8.7	7.2	7.7	7.0	6.3	9.2	5.2	6.6	6.7	6.8	11.0
6	8.8	7.4	7.8	7.2	6.6	9.3	5.5	6.8	7.0	7.1	11.1
7	8.9	7.5	7.8	6.9	6.3	9.5	5.4	6.7	6.9	7.1	11.3
8	8.6	7.3	7.7	7.1	6.4	9.4	5.3	6.6	6.9	6.9	11.1
9	8.5	7.1	7.8	7.2	6.7	9.6	5.4	6.8	7.2	7.4	11.4
10	8.3	7.1	7.8	7.0	6.5	9.3	5.2	6.6	6.8	6.9	11.2
12	8.4	7.1	8.0	7.3	6.7	9.5	5.4	6.8	7.0	7.2	11.3
13	8.4	7.0	7.8	7.1	6.8	9.4	5.2	6.7	6.9	7.3	11.4
17	8.5	7.1	7.8	7.3	6.8	9.6	5.3	6.6	7.1	7.2	11.4
18	8.5	7.0	7.7	7.4	6.9	9.9	5.3	6.6	7.3	7.6	11.4

	Labio-Lingual										
	Upper teeth						Lower teeth				
Village	I^1	I^2	C^1	P^2	M^2		I_1	C_1	P_1	P_2	M_1	N
1	7.2	6.4	8.4	9.3	11.6		6.0	7.3	7.6	8.2	10.8	9
2	7.1	6.3	8.0	9.7	11.4		5.8	7.1	7.6	8.3	10.5	18
3	7.3	6.3	8.2	9.8	11.6		5.9	7.2	7.9	8.5	10.8	56
4	7.3	6.5	8.2	9.6	11.5		5.9	7.1	7.8	8.5	10.8	26
5	7.4	6.4	8.2	9.6	11.3		5.8	7.2	7.9	8.4	10.7	47
6	7.4	6.5	8.3	9.7	11.5		6.1	7.4	7.9	8.5	10.6	38
7	7.3	6.7	8.5	10.0	11.8		6.2	7.5	8.1	8.5	10.8	24
8	7.4	6.6	8.2	9.7	11.3		5.9	7.3	7.8	8.3	10.6	42
9	7.2	6.5	8.1	9.8	11.7		6.0	7.5	7.9	8.5	10.9	29
10	7.3	6.5	8.3	9.9	11.6		6.2	7.5	8.0	8.4	10.7	64
12	7.5	6.7	8.5	10.2	11.8		6.3	7.7	8.2	8.7	10.9	22
13	7.4	6.7	8.4	9.7	11.7		6.1	7.4	8.0	8.3	10.8	16
17	7.3	6.6	8.4	9.8	11.7		6.1	7.5	8.0	8.5	10.8	20
18	7.2	6.5	8.2	9.8	11.7		6.2	7.4	8.2	8.6	10.8	58
												469

or labio-lingual diameters on different teeth within the same morphological class. The only correlation coefficients over .50 which involve an M-D and L-L combination is for the two diameters of the female M_1 and C^1. While males have 12 and females have 13 correlation coefficients above .50, only one of the female 13 involves an M-D pair (the upper and lower canine diameters), against only four for the males. High male correlations tend to involve anterior teeth, while high female correlations tend to involve labio-lingual diameters of the canines and cheek teeth.

The univariate F-ratios resulting from the comparison of the variance among and within villages for each tooth diameter are presented in Tables 7.3a and

Table 7.2
Dentometric Correlation Matrix
(Females upper right, males lower left)

Variable[a]	1	2	3	4	5	6	7	8	9	10
1.	1.00	.48	.45	.30	.38	.05	.48	.47	.35	.30
2.	.59	1.00	.36	.30	.21	.07	.35	.40	.26	.23
3.	.40	.37	1.00	.33	.33	.15	.42	.54	.38	.27
4.	.33	.39	.32	1.00	.43	.10	.27	.29	.44	.34
5.	.30	.27	.36	.38	1.00	.10	.34	.30	.46	.45
6.	.13	.15	.20	-.01	-.04	1.00	.12	.10	.15	.08
7.	.54	.36	.27	.22	.26	.14	1.00	.45	.39	.31
8.	.43	.42	.53	.33	.29	.19	.35	1.00	.47	.31
9.	.33	.37	.35	.51	.44	.03	.24	.41	1.00	.49
10.	.21	.17	.19	.45	.42	-.02	.15	.24	.46	1.00
11.	.41	.29	.35	.32	.32	.26	.32	.36	.41	.26
12.	.48	.41	.42	.31	.25	.23	.32	.41	.34	.22
13.	.29	.38	.28	.19	.19	.21	.18	.28	.32	.17
14.	.30	.32	.44	.25	.22	.23	.27	.43	.32	.16
15.	.26	.25	.37	.24	.36	.22	.22	.32	.36	.19
16.	.26	.26	.30	.20	.29	.30	.19	.30	.30	.25
17.	.30	.30	.23	.22	.18	.29	.26	.39	.34	.16
18.	.17	.21	.28	.15	.15	.26	.23	.36	.18	.11
19.	.30	.19	.25	.37	.30	.16	.29	.30	.42	.36
20.	.32	.33	.33	.36	.36	.18	.34	.38	.37	.32
21.	.37	.31	.41	.30	.29	.22	.30	.41	.42	.29

[a]Variables are listed in the same order as in Tables 7.1 and 7.3.

7.3b for males and females, respectively. It appears that in males, most teeth demonstrate significantly more variation among villages, as compared to variation within villages. The exceptions to this are the M-D diameter of the maxillary canine and second molar. The pattern for females is quite different; only 13 of the 21 teeth show significant differences (at the .05 level), although the single highest F-ratio is for the female P_2 M-D measurement. All of this suggests that females are more conservative phenotypically and do not express genetic variation in this set of measures so clearly as males. No purely demographic explanation, involving a higher frequency of female outmarriage, for example, is likely in light of what is known about current migration patterns.

In both males and females the same three measurements vary the most from village to village. These are all mesio-distal diameters of the premolars, measurements 4, 5, and 10. In males, a group of labio-lingual diameters on the anterior

Table 7.2 (continued)

	11	12	13	14	15	16	17	18	19	20	21
1.	.40	.40	.27	.36	.33	.37	.26	.27	.37	.39	.41
2.	.28	.30	.35	.29	.21	.27	.24	.24	.26	.19	.25
3.	.38	.40	.30	.51	.37	.38	.26	.39	.36	.36	.34
4.	.27	.21	.19	.21	.37	.30	.16	.20	.41	.36	.32
5.	.32	.23	.16	.22	.35	.27	.17	.17	.34	.42	.28
6.	.18	.17	.20	.23	.25	.33	.21	.23	.16	.18	.18
7.	.34	.27	.18	.32	.21	.27	.24	.21	.25	.25	.27
8.	.39	.34	.31	.41	.21	.31	.28	.31	.33	.32	.35
9.	.38	.23	.23	.25	.25	.30	.24	.21	.41	.42	.33
10.	.32	.16	.15	.16	.26	.32	.11	.10	.35	.38	.29
11.	1.00	.34	.27	.35	.29	.44	.25	.26	.34	.40	.56
12.	.37	1.00	.55	.58	.34	.43	.47	.44	.39	.41	.45
13.	.26	.61	1.00	.51	.35	.42	.52	.45	.36	.37	.35
14.	.31	.58	.57	1.00	.40	.44	.45	.56	.40	.41	.45
15.	.27	.38	.33	.41	1.00	.50	.31	.28	.43	.44	.35
16.	.34	.44	.41	.47	.44	1.00	.39	.36	.48	.53	.51
17.	.34	.54	.50	.47	.35	.41	1.00	.49	.38	.35	.37
18.	.14	.41	.36	.52	.24	.32	.46	1.00	.41	.35	.34
19.	.26	.36	.31	.34	.35	.38	.40	.34	1.00	.63	.49
20.	.30	.38	.34	.39	.36	.38	.30	.29	.51	1.00	.53
21.	.53	.39	.36	.44	.33	.41	.35	.31	.43	.53	1.00

Table 7.3a
F-Ratios for Male Dental Measurements

	Variable	F-Ratio	Probability
1.	I^1 M-D	2.92	0.67×10^{-3}
2.	I^2 M-D	3.01	0.49×10^{-3}
3.	C^1 M-D	1.52	0.11
4.	P^1 M-D	5.18	0.19×10^{-5}
5.	P^2 M-D	5.23	0.17×10^{-5}
6.	M^2 M-D	1.14	0.32
7.	I_1 M-D	2.50	0.30×10^{-2}
8.	C_1 M-D	2.03	0.17×10^{-1}
9.	P_1 M-D	3.10	0.37×10^{-3}
10.	P_2 M-D	4.36	0.11×10^{-4}
11.	M_1 M-D	2.43	0.39×10^{-2}
12.	I^1 L-L	3.02	0.47×10^{-3}
13.	I^2 L-L	3.86	0.38×10^{-4}
14.	C^1 L-L	3.81	0.44×10^{-4}
15.	P^2 L-L	2.10	0.13×10^{-1}
16.	M^2 L-L	2.91	0.68×10^{-3}
17.	I_1 L-L	3.69	0.60×10^{-4}
18.	C_1 L-L	2.82	0.93×10^{-3}
19.	P_1 L-L	3.46	0.12×10^{-3}
20.	P_2 L-L	2.03	0.17×10^{-1}
21.	M_1 L-L	2.29	0.65×10^{-2}

Table 7.3b
F-Ratios for Female Dental Measurements

	Variable	F-Ratio	Probability
1.	I^1 M-D	3.11	0.34×10^{-3}
2.	I^2 M-D	2.26	0.71×10^{-2}
3.	C^1 M-D	1.13	0.33
4.	P^1 M-D	4.89	0.31×10^{-5}
5.	P^2 M-D	4.64	0.53×10^{-5}
6.	M^2 M-D	1.81	0.39×10^{-1}
7.	I_1 M-D	1.16	0.31
8.	C_1 M-D	0.82	0.64
9.	P_1 M-D	3.94	0.29×10^{-4}
10.	P_2 M-D	5.55	0.86×10^{-6}
11.	M_1 M-D	3.16	0.29×10^{-3}
12.	I^1 L-L	0.78	0.68
13.	I^2 L-L	1.63	0.73×10^{-1}
14.	C^1 L-L	1.52	0.11
15.	P^2 L-L	1.50	0.11
16.	M^2 L-L	1.94	0.24×10^{-1}
17.	I_1 L-L	3.01	0.47×10^{-3}
18.	C_1 L-L	2.31	0.58×10^{-2}
19.	P_1 L-L	2.78	0.10×10^{-2}
20.	P_2 L-L	1.32	0.20
21.	M_1 L-L	2.07	0.15×10^{-1}

teeth are a second group of significantly varying measurements, but evidently these are not so important in distinguishing female characteristics from village to village.

Discriminant Analysis

A discriminant analysis was performed on the 21 dental variables for each sex separately, using the same approach reported in the anthropometric and dermatoglyphic sections. For both sexes, the H_1 test of the equality of dispersions was not significant, so that, unlike the dermatoglyphics, there can be no question about their suitability for discriminant analysis. As expected, the overall discrimination among males is better than among females, with a Wilks's Lambda of .300 for females, and .240 for males (Tables 7.4a and 7.4b). Of course, both are still highly significant, but less so than the male anthropometric results.

For both sexes, the first discriminant function is by far the most important, accounting for fully 41 percent of the discrimination in females and 30 percent in males. The fifth discriminant reaches significance in the males, while only the first three are significant statistically in the females.

The relationships among village centroids on the first three functions for males and females are presented in Figures 7.1 and 7.2 and for all significant functions in Tables 7.5a and 7.5b. The first function for both sexes separates the villages into two remarkably distinct groups, villages 1–8 and villages 9–18. Once again, the north-south dichotomy reappears. Judging from the discriminant loadings (Tables 7.6a and 7.6b), it is the same three premolar mesio-distal diameters which predominate in this function, with the northerners having the smaller teeth.

The second function for the male discrimination seems peculiar to it, stressing the labio-lingual diameters of I_1, I^2, C^1, and P^1, evidently accounting for the second group of significant measurements among males. In terms of separating out villages along this discriminant, it is hard to see any neat clusterings which are reminiscent of either demographic or other phenotypic relationships.

The other functions in both discriminant analyses account for relatively small portions of the variation and are increasingly difficult to interpret. Their discriminant loadings and village centroid scores are presented, nonetheless, in Tables 7.6a and 7.6b.

Table 7.4a
Relative Importance of Male Dentometric Discriminant Functions

Wilks's Lambda for total discrimination = .2401
Rao's F = 2.16
DF_1 = 273 DF_2 = 4298.4
Chance probability of F = .36 x 10^{-9}

Decreasing probability of significance of successive roots

Root	Eigenvalue	Percent of trace	χ^2	Probability
0			570.0	1.0 x 10^{-18}
1	.463	29.17	418.0	5.0 x 10^{-12}
2	.267	16.79	323.6	.656 x 10^{-6}
3	.188	11.83	254.8	.205 x 10^{-3}
4	.164	10.32	194.2	.136 x 10^{-1}
5	.126	7.94	146.8	.123

Note: Chi-square tests of significance with successive roots removed.

Test of H_1 (equality of dispersions)
  F = .760; associated probability = 1.00
  DF_1 = 3003, DF_2 = 67676

Table 7.4b
Relative Importance of Female Dentometric Discriminant Functions

Wilks's Lambda for total discrimination = .3005
Rao's F = 2.039
DF_1 = 273 DF_2 = 4864.5
Chance probability of F = .12 x 10^{-8}

Decreasing probability of significance of successive roots

Root	Eigenvalue	Percent of trace	χ^2	Probability
0			541.7	1.0 x 10^{-17}
1	.559	41.10	341.6	.176 x 10^{-4}
2	.184	13.52	265.5	.495 x 10^{-2}
3	.170	12.50	194.7	.215
4	.107	7.87	148.9	.579
5	.092	6.78		

Note: Chi-square tests of significance with successive roots removed.

Test of H_1 (equality of dispersions)
  F = 1.021, associated probability = .219
  DF_1 = 3003, DF_2 = 29311

7.1   Male dentition: centroids of the 14 samples on the first three discriminant axes

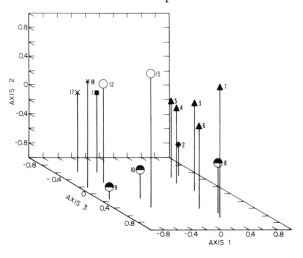

7.2   Female dentition: centroids of the 14 samples on the first three discriminant axes

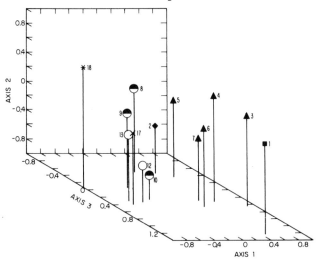

Table 7.5a
Male Dentometrics—Group Centroids
in Discriminant Space

Village	Function				
	1	2	3	4	5
1	.38	.09	-.61	.25	-.39
2	.72	-.57	-.50	.52	.12
3	.57	.21	-.02	-.78	.10
4	.50	.05	-.28	-.18	-.90
5	.56	-.03	-.46	.15	.19
6	.20	.15	.43	.20	.54
7	.28	.82	.65	.35	.15
8	.38	-.30	.54	.13	-.20
9	-.84	-.83	.17	.09	-.12
10	-.40	-.60	.15	-.28	.07
12	-.53	.38	-.31	-.08	.32
13	-.44	.85	.40	.60	-.69
17	-.63	.07	-.63	.58	.19
18	-.88	.47	-.18	-.22	-.08

Table 7.5b
Female Dentometrics—Group Centroids
in Discriminant Space

Village	Functions		
	1	2	3
1	.44	.26	1.26
2	.39	-.34	-.47
3	.89	.25	.44
4	.53	.45	.35
5	.56	.05	-.35
6	.23	.07	.48
7	.35	-.13	-.30
8	.14	-.15	-.48
9	-.39	.04	-.01
10	-.32	-.68	.25
12	-.51	-.50	.36
13	-.66	-.08	.32
17	-.63	-.01	.04
18	-1.00	.65	-.07

Table 7.6a
Discriminant Loadings—Male

Variable	1	2	3	4	5
1. I^1 M-D	-.14	.42	.32	-.25	.16
2. I^2 M-D	-.03	.34	.50	-.06	.01
3. C^1 M-D	-.19	.32	-.04	-.04	-.09
4. P^1 M-D	-.51	.44	-.03	-.05	.03
5. P^2 M-D	-.57	.37	.10	-.11	.09
6. M^2 M-D	-.14	.26	-.07	-.11	-.10
7. I_1 M-D	-.04	.22	.54	.16	.00
8. C_1 M-D	-.18	.30	.30	-.23	.04
9. P_1 M-D	-.38	.37	.01	-.15	-.07
10. P_2 M-D	-.46	.38	-.06	.17	.15
11. M_1 M-D	-.31	.20	.15	-.24	-.05
12. I^1 L-L	-.08	.17	.42	-.43	.40
13. I^2 L-L	-.10	.52	.48	-.14	.17
14. C^1 L-L	-.03	.51	.12	-.20	.50
15. P^2 L-L	-.23	.16	.07	-.21	.42
16. M^2 L-L	-.17	.37	.13	-.48	-.05
17. I_1 L-L	-.41	-.04	.40	-.22	.26
18. C_1 L-L	-.32	.13	.21	-.27	.06
19. P_1 L-L	-.22	.49	.19	-.40	.01
20. P_2 L-L	-.10	.36	-.08	-.34	.13
21. M_1 L-L	-.14	.25	.12	-.54	-.07

Table 7.6b
Discriminant Loadings—Female

Variable	1	2	3
1. I^1 M-D	.30	.35	-.28
2. I^2 M-D	.07	-.21	-.48
3. C^1 M-D	.08	-.03	.30
4. P^1 M-D	-.53	.16	-.26
5. P^2 M-D	-.53	.17	.03
6. M^2 M-D	-.17	.33	.20
7. I_1 M-D	-.05	.03	-.23
8. C_1 M-D	.04	-.02	-.03
9. P_1 M-D	-.43	.34	-.11
10. P_2 M-D	-.52	.39	-.11
11. M_1 M-D	-.39	.01	-.06
12. I^1 L-L	.06	-.08	-.01
13. I^2 L-L	-.20	-.12	-.10
14. C^1 L-L	-.08	-.17	.23
15. P^2 L-L	-.13	-.23	.07
16. M^2 L-L	-.18	.00	.27
17. I_1 L-L	-.37	-.16	.19
18. C_1 L-L	-.29	-.30	.09
19. P_1 L-L	-.28	.15	.11
20. P_2 L-L	-.12	.16	.08
21. M_1 L-L	-.12	.18	.29

Generally, then, the dental variables do reflect a major split among the sample villages, but within these subgroups, variation is hard to interpret. The premolars are by far the most important teeth in describing what variation there is, with some appreciably significant variation in the anterior teeth apparent in males.

Using the centroid rule, the percentage of correct assignments of individuals to different groups is .357 for males—a relatively low value compared to the discriminatory powers of the anthropometrics but higher than that of the dermatoglyphics (Table 7.7). The percentage of correct assignments in the female sample is in the same range, worse than the anthropometrics but better than the dermatoglyphics.

In conclusion, the gross dental measurements in this series, taken three years after the initial survey, show the same marked division between the northern Papuan-speaking population and the southerners that is evident in the anthropometrics and blood polymorphisms. While female variation along this north-

Table 7.7
"Hits" and "Misses"—Male[a]

	1	2	3	4	5	6	7	8	9	10	11	12	13	14
1.	3	4	1	1	2	0	1	0	1	0	0	1	1	0
2.	3	10	2	0	2	0	0	0	0	1	0	0	2	0
3.	2	1	14	2	5	2	0	3	2	3	1	2	3	3
4.	4	2	2	8	0	0	0	2	0	2	0	1	0	1
5.	1	5	1	6	14	2	3	1	0	2	2	2	1	1
6.	3	1	4	2	3	8	4	2	1	1	1	2	1	1
7.	1	2	3	1	0	2	7	2	0	0	1	2	0	1
8.	1	4	5	6	1	3	1	12	6	0	2	3	3	0
9.	0	1	0	2	1	1	2	1	15	2	0	0	0	4
10.	1	1	3	4	2	3	1	1	6	14	3	2	2	2
11.	1	1	2	0	2	0	2	2	1	1	7	2	5	2
12.	0	1	0	0	0	1	2	0	1	0	1	7	1	0
13.	1	0	0	0	0	0	2	0	2	1	2	0	9	0
14.	3	1	0	2	0	0	1	2	2	4	4	0	1	21

[a]Row is actual group membership; column is membership by identification rule
Total Hits = 149 out of 418. Percentage = .3565

south axis is quite marked, total male variation is appreciably greater, suggesting the developmental conservatism of females.

This discriminatory power of such a simple dental exercise is remarkably good and should encourage anthropologists and geneticists to include dental casting in their standard test battery.

How do the different patterns of variation described in the earlier sections compare with one another and with the language and geographic relations of the villages? This is a particularly thorny problem, involving as it does an evaluation of patterns which are already about three steps removed from the realities of individual measurements or tests on individual subjects or samples.

Currently in the literature are three or four different approaches to this general problem. One involves correlations of the distances between pairs of groups with respect to the different sorts of biological variation. For example, where blood group, anthropometric, and geographic relationships are known among four tribes, there will be six "distances" among the paired groups (1-2, 1-3, 1-4, etc.) in relation to each of the three different sorts of variation. These three sets of six distances can be correlated with one another to derive a measure of association among anthropometrics, blood polymorphisms, and geography (see, for example, Friedlaender et al. 1971, Howells 1966a, Hiernaux 1956). The problem is that the correlation coefficient is a somewhat labile statistic and assumes a number of conditions which such pairwise distance comparisons violate.

In the first place, separate observations which are to be correlated should be independent of one another. This is patently not the case, as once geographic distances between tribe 1 and the others are known, the distances between any of the other 2 tribes are limited and become more constrained as more distances are established.

Secondly, the distribution of the observations in all sets should be normal for any test of significance to be valid. For most of the sets in the Bougainville series, this requirement is not met because there is usually an excess of shorter distances.

As a result, correlation coefficients are an unhappy approximation to a useful statistic in this case. Nonparametric measures of correspondence, which have fewer such assumptions, such as Kendall's Tau, are probably more appropriate, but have less power than strict parametric statistics. However, it is Kendall's Tau which will be used at a later stage as a corroborating test of association.

Recently, some authors have attempted another comparative technique (see Ward 1972, Ward and Neel 1970). This involves deriving branching diagrams, or trees, from matrices of distances among a set of samples, and then comparing the trees from different biological sources. Statements resulting from such comparisons are to the effect that "the best tree representing the anthropometric distance relationships of the groups is among the best 1 percent of the possible trees which represent the blood polymorphism relationships," or something of the kind. The main reservation to this approach is that in most comparisons there are so many possible trees that percentage statements have little meaning, especially as the nature of the distribution of likely trees is probably either unknowable or unique for each new case. Secondly, and more importantly, the comparison involves phenomena (trees) which are even further abstracted from the data and hence less reliable reflections of reality.

As a footnote, studies of "population structure," so called by Newton Morton (see Morton 1968, Imaizumi et al. 1971, Langaney and Gomila 1973, and also Friedlaender 1971a, b), are closely related to population distance statistics because they are tied to measures of the genetic variance, as are distance statistics (see Appendix). However, there are problems in trying to derive population structure statistics for traits such as dentometrics, anthropometrics, and dermatoglyphics, where the genetic basis of the variation is not clear (Friedlaender 1974). Also, there are problems in such studies in comparing results from different sample universes, as the genetic variance is so heavily dependent upon the sampling size (Friedlaender 1974). Therefore, comparisons of population structure statistics from Bougainville and American cities, for example, are probably inappropriate, as the nature of the breeding populations in the two areas is completely different.

A happier solution to the entire problem of comparing biological and demographic patterns from the same populations is one which has been suggested by John Gower (1966 and 1971) for comparing metric results taken on different parts of the cranium (see also the Appendix at the end of this book). His approach is relatively simple.

As an example, if the different patterns of the village relationships are diagrammed on transparent sheets (so that the anthropometric relations on discriminants 1 and 2, presented in Figure 5.9 would be one such diagram, as would the geographic map of the sample area), then by superimposing the different diagrams on one another and by shifting the graphs around until the villages on the sheets are in the closest corresponding positions, then it is possible to decide which diagrams are most alike in their patterns, and which are less so. It should also be legal to reflect the graph axes, or turn the graph sheets over, in order to get the best approximation. In previous years such a procedure was called a perversion. To put a number on it, the "fit" of any two diagrams can be measured by summing up the distances between corresponding points on the graphs (or their deviations) once the best rotations and perversions have been made.

It makes sense that corrections for differences in scale between graphs or sets of data should be allowed, so that deviations in gene frequencies may be compared with map distances in kilometers, for example. Also, multidimensional comparisons should be feasible.

The better the fit between two arrays, the smaller the summed deviations (squared) should be. Gower has called this statistic R^2, and at this point very little is known about its distribution, so that significance tests are impossible. Comparisons in graphical terms are presented in later sections of this chapter. Workman et al. (1973) have used essentially this technique to compare blood polymorphism, migration, and geographical relationships among the Papago Indians, finding that past migration rates correspond best with current genetic relationships as revealed by blood polymorphisms.

This is not to say that Gower's approach is without fault. It is entirely plausible to ask whether or not uniform scaling of different sets of "distances" is in fact possible or even advisable. Along a single dimension, should one, for instance, attempt to compare squared deviations or distances (such as Mahalanobis' D^2) with unsquared geographical distances or migration-relationship coefficients, or is it more appropriate to transform such distance statistics? Another problem involves the comparison of patterns which have different dimensionalities. All other things being equal, a two- or three-dimensional swarm of points, such as the blood polymorphism array seems to be, should be more closely associable with geographic relationships than should the anthropometric results, which are properly presented in at least four dimensions. But, comparing swarms of points in different dimensions is also a problem for pairwise distance correlations, although obviously less so. Ascertaining levels of significance for these statistics is a particularly intractable problem, and generating "random" sets for comparative purposes is not an attractive alternative.

One approach that might be highly instructive is a nested multivariate analysis of variance of the different sets of data to establish the relative degree of variation among villages within language groups, as opposed to that among language groups within the sample. This is essentially analogous to the use of Wahlund's f in Chapters 3 and 4.

Results

For the comparisons of the different arrays it is necessary to eliminate those samples from consideration that are not represented in all configurations. This reduces the number of villages to 14 (limited by the dental analyses), excluding the Siwai groups 14–16, and the tiny village of Sieronji in North Nasioi (11). Moreover, while the villages may be more or less the same, the actual individual compositions of the samples do not overlap completely. The anthropometrics were taken on adult males residing in those villages in 1967; the dental casts were taken three years later on individuals who had been included in the previous sample in some capacity and who were still residing in the area; the blood samples, a pooled male and female sample, include everyone over the age of three in residence; and the finger and hand prints include essentially everyone over the age of ten. At best, then, the array represents repeated sampling of the same genetic universe with replacement. Solely on the basis of sampling similarities, one would expect the fingerprint and blood polymorphism results to bear the closest resemblance, with anthropometrics having a substantial relationship to male fingerprints and secondarily male dentometrics, and likewise to female dermatoglyphics and dental measurements. This does not work out.

To be certain that there were no major differences from the original descriptions and the results of the 14 population samples, new discriminant analyses were performed. The same configurations and the same sort of discriminant loadings seemed inevitably to present themselves. Judging from this set of village discriminants (see Table 8.1), the discrimination is most successful for the anthropometrics, secondly for the dental measurements, and least significant for the male and female dermatoglyphics.

Turning to the Gower approach, Table 8.2 presents the R^2 comparisons of the patterns derived from the nine different sources—geography, language,

Table 8.1
Wilks's Lambda for the Discriminant
Analyses of 14 Villages

	Λ	Rao's F	NDF_1	NDF_2	Probability
Anthropometry	.1585	5.03	169	3866	$.16 \times 10^{-13}$
Male dentometrics	.2401	2.16	273	4298	$.36 \times 10^{-9}$
Female dentometrics	.3005	2.04	273	4864	$.12 \times 10^{-8}$
Male dermatoglyphics	.4145	1.37	364	6463	$.30 \times 10^{-4}$
Female dermatoglyphics	.4243	1.42	364	6864	$.83 \times 10^{-5}$

Table 8.2
Gower's R^2 Distance among Bougainville Patterns

	Anthropometry 1	Male teeth 2	Female teeth 3	Male fingers 4	Female fingers 5	Blood 6	Geography 7	Language 8	Migration 9
1.									
2.	.57								
3.	.59	.47							
4.	.99	.78	.90						
5.	1.02	.97	.91	.85					
6.	.53	.75	.70	.74	.77				
7.	.64	.82	.64	1.17	.84	.53			
8.	.34	.76	.61	1.08	.99	.55	.59		
9.	.56	.80	.64	.66	.78	.63	.84	.77	

migration rates, blood polymorphisms, male and female hand prints, male and female tooth measurements, and male anthropometrics. The number of dimensions needed to represent the significant variation in each case was: geography, 2; blood polymorphisms, 3; female dentition, 3; language, 4; migration, 4; male dermatoglyphics, 4; anthropometrics, 5; male dentition, 5; and female dermatoglyphics, 5.[1]

1. One last quirk to note is that all the patterns have 14 points, or groups, to compare, except for the language relationships, where there are effectively only 7 different points, or language groups, represented. Villages belonging to the same language group were assigned the same point in space.

Remembering that the smaller the value of R^2 the better the correspondence, we find that the anthropometry and language patterns are by far the closest, with a value of .34. The congruence of the two patterns on their first two axes are presented in Figure 8.1, with the language points named and the corresponding anthropometric points numbered. The black connecting lines are the deviation lines, or what is measured by R^2, and are reasonably short. Note that the fit between the two sets is particularly good for the northern areas—the Aita, Rotokas, Eivo, and Simeku—but less good for the Nasioi- and Melanesian-speakers nearest Kieta township, where the linguistic relations appear to be substantially more disparate than the anthropometric ones. It is impossible to resist recalling that there has been substantial recent intermarriage among these more southerly groups, and that the Uruava, village 17, are in fact predomi-

8.1 Comparison of language group and anthropometric relationships among 14 villages

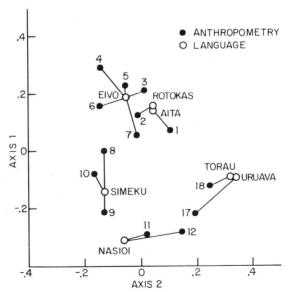

nately Nasioi-speakers today. In this case, language relationships are changing more slowly than physical ones.

That the migrational relationships are best represented in more than two dimensions may be surprising at first, but realize that the uneven dispersion of marital distances, which is fairly well approximated by a negative exponential curve, will distort predicted relationships into more than two or three dimensions. If an appropriate transformation were made, the dimensionality might be reduced.

The next three "best" fits are between anthropometry and blood polymorphisms, blood polymorphisms and geography, and geography and language. Figure 8.2 presents one of these good comparisons, between geography and blood, as a further illustration of the method. At least on the first two functions,

8.2 Comparison of geographic and blood polymorphism relationships among 14 villages

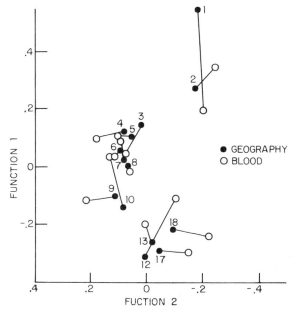

the correspondence of points is generally quite good, with the major deviations involving villages 1, 10, and 12.

The distinctly poor correspondences involve the male and female dermato-glyphics. It is surprising, though, to find that their closest correspondences are not even with one another, but with migration (for male dermatoglyphs) and blood polymorphisms (for female dermatoglyphs). As a contrast to the two other graphs, the correlation between female dermatoglyphs and geography is presented in Figure 8.3. The maze of deviation lines defies analysis.

In an attempt to simplify the table of R^2 values, and to make more sense of it, the principal coordinates of this matrix of "distances" were computed to display these distance relationships in simplest geometric space. Admittedly,

8.3   Comparison of geographic and female dermatoglyphic relationships among 14 villages

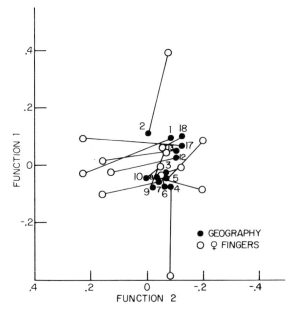

this is an abstraction to yet another level. Nevertheless, Table 8.3 and Figures 8.4 and 8.5 present the results.

Inevitably, the dermatoglyphics are furthest removed measures from the rest, with predicted migration patterns only somewhat better. The closest correspondences are between anthropometry and language and between male and female teeth, with blood, language, and geography forming a rather cohesive group.

As a corroborating test, Kendall's Tau statistic was calculated for these various "distance" pairs. The results, and the associated significance tests, are presented in Tables 8.4 and 8.5. Differences in results from R^2 are to be expected; for example, the language relationships must be represented in this calculation as "ties." On Kendall's Tau, geography and language are closely re-

8.4 Congruence of the 9 patterns of variation over the first and second dimensions

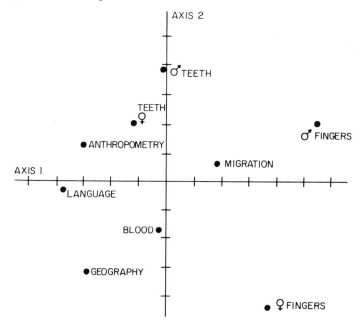

Table 8.3
Principal Coordinates Analysis of
Gower's R^2 Relations among 9 Variables

Pattern	Axis 1	Axis 2	Axis 3
Anthropometry	−.298	.128	−.202
Male teeth	−.014	.376	.315
Female teeth	−.117	.203	.281
Male fingers	.551	.199	−.145
Female fingers	.374	−.441	.223
Blood	−.033	−.172	−.203
Geography	−.285	−.323	.145
Language	−.367	−.035	−.156
Migration	.190	.063	−.259

8.5   Congruence of the 9 patterns of variation over the first and third dimensions

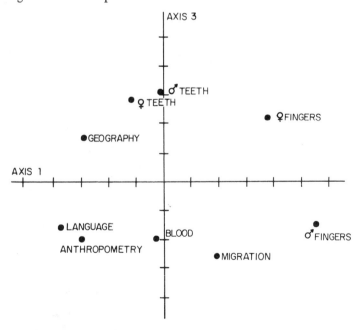

Table 8.4
Measures of Congruence of
Bougainville Patterns, Kendall's Tau

	Anthropometry 1	Male teeth 2	Female teeth 3	Male fingers 4	Female fingers 5	Blood 6	Geography 7	Language 8	Migration 9
1.	1.00								
2.	.38	1.00							
3.	.35	.33	1.00						
4.	.12	.17	.13	1.00					
5.	.05	.01	.07	.39	1.00				
6.	.31	.19	.34	.29	.22	1.00			
7.	.44	.25	.53	.26	.27	.54	1.00		
8.	.42	.24	.41	.08	.08	.42	.50	1.00	
9.	.25	.25	.26	.35	.29	.34	.36	.22	1.00

Table 8.5
Significance of Kendall's Tau in Normal Deviates

	Anthropometry 1	Male teeth 2	Female teeth 3	Male fingers 4	Female fingers 5	Blood 6	Geography 7	Language 8	Migration 9
1.									
2.	5.3								
3.	5.0	4.7							
4.	1.7	2.4	1.8						
5.	.7	.2	1.0	5.5					
6.	4.4	2.7	4.8	4.1	3.1				
7.	6.2	3.5	7.5	3.6	3.9	7.6			
8.	5.9	3.3	5.8	1.1	1.1	5.8	7.1		
9.	3.4	3.6	3.7	4.8	4.1	4.8	5.1	3.1	

lated to each other, and to anthropometry, blood polymorphisms, and female dentition, a newcomer to the fold. However, most of the variables are significantly related at rather elevated levels, with the exception of the dermatoglyphic comparisons. Most of the patterns are reasonably close approximations to one another, and there is little to choose among the best two or three.

Conclusion

What does it all mean? To hark back to the Introduction, this small area of Bougainville has certainly proved to be unexpectedly rich in human biological diversity. This does not mean that all aspects of the phenotype show the same degree of variation, or even vary at all from one village to the next. The finger and hand prints are excellent examples of traits which differ little among the villages, but which vary enormously from one individual to the next. In contrast, many traits, such as skin color, hair form, and eye color, are monotonously uniform over the area.

This relates to a classic interest of physical anthropology, the elucidation of different sorts of biological variability within and among human populations. In this survey, it has proven relatively easy to tie *average patterns* of variation among the village samples to essentially demographic determinants. The more an aspect of the phenotype does vary among the villages (as judged by Wilks's Lambda in Table 8.1), the closer the correspondence of the resulting biological pattern with the predicted relationships from migration, language, and geographic information. This suggests that the differences in anthropometry and dental measures are hereditary, as it is to be expected that the effects of population subdivision will be most clear-cut in those aspects of the phenotype with the highest heritability. However, dermatoglyphic variation, which has been assumed to have a high heritability, appears to be a relatively insensitive indicator of variation among populations. It may be that the relatively enormous dermatoglyphic variation from individual to individual effectively obliterates or obscures the differences among groups.

Average genetic differences among the groups are not determined by their adaptations to different ecological and environmental situations (embodied as agents of selection or phenotypic "modifiers"), which would have different effects on different organ systems. Selection is undoubtedly at work in these populations, but its effects, which act to differentiate the populations, are ignored or averaged out. Over larger continental areas, where great cultural and ecological differences should intrude, it is unlikely that the same explanations can persist.

It is unlikely, in fact, that studies of this sort can be expected to yield similarly clear results in many other areas of the globe. Groups of socially unstratified, sedentary people whose relationships have been relatively stable for centuries or millennia are on the wane. Only the side streams and backwaters of the Heraclitan river of humanity have been relatively unaffected by the flood of more people and new technologies in this century. Most of the current variation will undoubtedly persist, but the influences of its origins and sources will become more difficult to recognize.

If this is true for the village and small language area in Melanesia, it is that much more certain for larger, more inclusive groupings of humans. People will inevitably persist in the naming of racial groups based on simple physical and even social attributes, but, at the very least, they should be made aware how grossly simplified any such taxonomic system has to be, and the diversity which a name such as "Melanesian" masks.

Appendix
Bibliography
Index

Appendix
Techniques of Data Collection and Analysis

I. Field Procedures
The collection of data in each of the villages followed a set routine, both in the initial (1966–67) and final (1970) surveys.

Demography and Genealogy
If there were adequate demographic records available at the local mission stations (true for the east coast Manetai and Tunuru parishes), these were transcribed onto 3 × 5 cards, one for each biological nuclear family, with dates of birth included for all individuals where this was established or estimated.

In the village I would begin as soon as possible to interview and collect data from family groupings. My assistant and the village counsellor or head man ordinarily suggested which families should come in which order and at what particular time. I would locate the family record card (if it existed), check the information on it as closely as possible, and also ask adults their places of birth, their previous marriages and offspring by them, the names and residences of their siblings, their parents and their birthplaces, and the number of children (alive and dead) by their current marriage. Survey numbers were assigned to each person included in at least one of the data series other than the census.

Toward the end of the village stay, when all individuals who were likely to be included had been initially covered, I constructed a genealogical chart for the living members of the village in which I used the information I had previously gathered, and checking data with older men and women of the village. Although the extent of the genealogy varied, it usually covered the last three generations, with some matrilineal and patrilineal descendants being traced back substantially further.

In 1970 I attempted to connect current genealogical records of my own in northeast Siwai with those Douglas Oliver constructed in 1938 and 1939. This procedure did not appreciably extend the genealogical record, but served to fill out the information on the last three generations and to corroborate the validity of my own records.

Anthropometry
After the initial family census, I measured the adult male(s) of that family. I chose a short series of fifteen measures and weights, based largely upon the series that constitutes Penrose's size and shape descriptions (1954). The measurement definitions follow:
Body Measurements
Weight. Measured in pounds with a "Detecto" spring scale like that used by the

Harvard Solomon Islands Expedition. Weight calibration was checked roughly at the beginning of each village survey with 80 pounds of equipment.

Stature or Standing Height. The distance from the highest point of the top of the head in the midsagittal plane to the floor. Taken with anthropometer with wood board as standing platform. Subject standing at attention.

Sitting (Vertex) Height. The subject is seated on a low "patrol box," with feet on the ground, so that his ischial tuberosities are in close approximation to the surface of the box. The subject is prodded in the midback into his most erect posture. The measurement is then taken with the anthropometer from the highest point of the head, in the sagittal plane, to the surface upon which the subject is seated (the patrol box).

Chest Breadth. (Taken with anthropometer as sliding caliper). The transverse distance between the most lateral points on the chest at nipple level. Subject in Standard Erect Position.

Chest Depth.[1] (With anthropometer as sliding caliper). Taken at nipple level, parallel with floor, and just below the inferior angle of the scapulae. Subject in Standard Erect Position.

Biacromial Breadth.[1] (Taken with anthropometer as sliding caliper). The distance between the most lateral margins of the acromion processes of the scapulae, the subject standing in relaxed posture.

Biiliac Breadth.[1] (Anthropometer used as sliding caliper). Breadth from iliocristale, the most lateral point on the crest of the ilium, to iliocristale.

Total Upper Extremity Length. (Anthropometer as sliding caliper). From acromiale (lateral margin of the acromial process of the scapula) to dactylion (the tip of the middle finger), when arm is hanging down, and palm faces inward. Taken on left side of body unless deformed.

Head Measurements

Head Length. (Spreading caliper). The distance between glabella (the most prominent point between the eyebrows, in the midsagittal plane) and the farthest projecting point in the midsagittal plane on the back of the head (occiput). This latter point is termed opisthocranion.

Head Breadth. (Spreading caliper). The greatest transverse diameter of the head, ordinarily found at points overlying the parietal bones (the point is called euryon).

1. These measurements taken only on older men who had been previously measured by Douglas Oliver in 1938–39.

Minimum Frontal Breadth. (Spreading caliper). The shortest distance between the temporal lines, taken above the orbits.

Bizygomatic Breadth. (Spreading caliper). The distance between the most laterally situated points on the zygomatic arches.

Bigonial Breadth. (Spreading caliper). The distance between the gonial points (the most lateral point upon the postero-inferior angle of the mandible.)

Total Face Height. (Sliding caliper). From nasion (the point at which a horizontal, tangential to the highest points on the superior palpebral sulci, intersects the midsagittal plane) to gnathion (the lowest median point on the lower border of the mandible). When a subject had no teeth, I noted the condition and made a "best estimated measurement."

Nose Length. (Sliding caliper). From nasion to subnasale (the point at which the nasal septum, between the nostrils, merges with the upper cutaneous lip in the midsagittal plane).

Nasal Breadth. (Sliding caliper). The maximum transverse distance between the most laterally situated points on the wings of the nose.

Dermatoglyphics

Prints of the palm and each finger were taken using Faurot printer's ink and ordinary airmail stationery. All resident adults and children were ordinarily printed, and the dermatoglyphic series is the largest and most complete as a result. However, for the purposes of this current analysis, it seemed sufficient to include only those individuals over ten years old who had no "missing" fingerprints (this included those prints so heavily scarred they could not be read).

Other Morphological Data and Observations Gathered, but Unreported

Anthroposcopic Observations. A series of fourteen observations (brow ridge thickness, forehead slope, occipital protrusion, Darwin's point, nasal tip inclination, integumental lip thickness, hair form, hair color, eye color, eye-opening height, skin color, third molars erupted, nasion depression) were made on the adult males, and have been analyzed. The results have not been included here however, because (a) the variation from one group to the next is minimal and (b) because what variation there seemed to be was most explicable in terms of observational "drift." I seemed to change my criteria as the months proceeded.

Visual Acuity. Tests for visual acuity, using an "E" chart and a cardboard cut-out E for the subject to indicate position with, were performed on all adults and adolescents. Perception was uniformly excellent, except in the cases of a few elderly men and women. There was no appreciable variation among the village samples.

Genetic Observations

Color Blindness. All male subjects who were able to comprehend the test (effectively those over 5 years) were tested for color blindness using the Ishihara charts (1967) for nonliterate people. Red-green color blindness was present in persistent levels in most of the language areas, sometimes approaching a frequency of 15 percent. A listing of the results by language group are presented in Table A.1. These results might have been included in the main text, but because of the small samples, I chose to exclude them. It is worth noting that this finding of color-blind individuals is in contrast to the results of the Harvard Solomon Island survey, and to ideas on the lack of color blindness in nonurbanized societies (Post 1962).

Phenylthiocarbamide Taste Sensitivity (see Kalmus 1958 for a basic review). I used an abbreviated series of taste tests, with two paper cups filled with PTC solution at the antimode of taste sensitivity and two others without, on all individuals over the age of four years. There was a low incidence of non-tasters, so that while this may be of some interest in larger surveys of the Pacific or East Asian populations, it has little significance in this book.

Blood Sampling

On the last day of my stay in each village, I took 10cc of venous blood from the antecubital veins of all individuals who had been covered in the earlier sections of

Table A.1
Incidence of Color Blindness in Bougainville Males

Language group	Color-blind individuals	N	Percent
Rotokas (and Aita)	0	49	0
Eivo	16	289	5.5
Simeku	17	200	8.5
Nasioi	17	118	14.0
Torau (and Arawa)	3	137	2.2
Siwai	5	94	5.3
Total	58	887	6.5

the survey—either in the census or the data-gathering phase—who were available and who were older than three years of age. Becton-Dickenson "Vacutainers" with no additives were used. The samples remained at room temperature for at least eight hours, allowing for maximum retraction of the red cells. Ordinarily the next morning they were cooled down to approximately 4° C in thermos flasks with frozen picnic tins. I then transported the samples to the nearest airstrip (Aropa, Buin, or Wakunai), where the bloods were flown out to Sydney via stops in Rabaul and Port Moresby, within the (then) Territory of New Guinea. Red cell antigens were tested within the first two days of arrival in R. J. Walsh's Laboratory, under the direction of P. Clark, and serum proteins were typed within two weeks. Approximately 1 percent of the samples hemolyzed, and some of the sera were bacteriologically contaminated by the time of their actual analysis. It is my belief that the hemolysis was largely due to the shaking of the samples once they were no longer properly refrigerated in transit to Australia.

Dental Materials

Because the sample populations were so rich in diversity, a second expedition was planned for Bougainville in July and August of 1970, primarily to obtain dental impressions to establish another pattern of variation for comparison. Dr. Howard Bailit, of the Anthropology and Dental Schools of the University of Connecticut, and his assistant Glen Rapoport, returned with me to Bougainville. While I concentrated on the genealogical reconstructions in Siwai, they proceeded to cover the fourteen villages on the eastern coast of the island in extremely fast and efficient fashion. They tried to obtain dental impressions (in an alginate material) on all residents present in a village during the day or so they resided in each village. Casts of dental stone were made immediately after the impressions were taken. Children who did not have at least four erupted permanent teeth were excluded from the study, as they would provide limited information for family comparisons—a major objective of the project. This means that most children below the age of six were not studied. For essentially the same reasons, elderly individuals with many permanent teeth missing or loose were not included in the sample. The only other villagers who were under-represented in the dental sample were young adult men working as contract laborers on coconut plantations or a nearby mining operation.

Three years had elapsed since I had collected the serological, fingerprint, and anthropometric data on the study population, and a question therefore arose as to the overlap with the dental sample; that is, how many individuals were sampled by both Bailit in 1970 and me in 1967. Of the 1200 people on whom dental casts were taken, 846, or 70 percent, also had been seen previously. Only these 846 casts, from 418 males and 428 females, comprise the dental sample for the current study.

The mesio-distal and labio-lingual diameters of the permanent teeth on the left side of the maxilla and mandible were measured from stone dental casts. In cases where a tooth was missing on the left side, its antimere on the right side was used in its place. Teeth which were excessively worn, chipped, or carious were not included in the sample.

II. Data Preparation and Analysis

Uniformity in analysis was the major goal, as it was in collection techniques.

Anthropometrics

All measurements and observations were punched and verified on Hollerith cards. As multivariate statistics of the sort performed on the anthropometrics, dentometrics, and dermatoglyphics require that all variables must be scored for each subject, careful estimation of a missing measurement can save the information carried in all the other measurements on that individual, as Howells (1973) points out.

In addition, there will be those measurements which, while accurate, will act as "errors" in obscuring normal biological variation, such as arm length on subjects with withered arms, or nose length on men with stretched nasal septa owing to the ornamental wearing of nose bones. Recording and measuring errors necessarily become more devastating in this sort of analysis, so that the entire problem of errors and omissions becomes critical.

Several techniques were used to deal with the situation. First, while in the process of measuring the individuals, notes were made when an unusual measurement was taken. If the abnormality was due to disease, accident, or ornamental design, an additional attempt at estimation of what the "normal" measurement might have been was made beside the actual one.

Second, all the anthropometrics and dentometrics were checked by a computer program "Editing Routine" written by E. Churchill of Antioch College (Churchill 1963) which also estimates values for missing data by regression.[2] The program

2. This is superior to the technique of inserting the mean figure for a series in a missing spot or for an obvious error. A small person might be given a disproportionately large value for a missing variable, distorting his total numerical configuration. Questionable figures or undetected recording or punching errors are located by the same procedure.

computes regression equations for predicting values of any measurement from two others highly correlated with it. It then reads the actual figures again, printing out all cases in which the deviation from prediction exceeds three times the standard error of estimate. As the triple combinations of measurements may be numerous, and their various values and predictions are printed out in each case, it is usually easy to establish whether the value is in fact an error or is simply a valid large deviation from the usual "shape" of the sample.

The embarrassingly long list of measurements which were printed out were checked against the original anthropometric and dentometric forms for recording and reading errors. A number of anthropometric errors of this sort were found, especially number reversals. Even after this examination there were still about 30 measurements which were questionable—that is, it was unclear that they were errors, not having been noted as particularly aberrant, and not lying far beyond the third standard error of estimate. I revised such values toward their predicted score, but only so that they did not deviate so markedly as they had. This was a way of suspending judgment, admittedly unsatisfactory, but it was hoped, unimportant enough to contribute more to the total mass of data than it subtracted.

Dentometrics

The measurements were taken by three people at the University of Connecticut at Storrs; Nancy Learned, a research assistant; Howard Blessing, a dental student; and Howard Bailit. Most of the measurements were obtained using an automated caliper attached to a digital printer. This procedure substantially reduces the time and probably the errors of tooth measurements (Conneally et al. 1968).

Before taking the measurements, the three examiners were calibrated with each other. In addition, duplicate measurements were obtained on a small sample of teeth, indicating that the intra- and inter-examiner errors were within limits reported by other investigators (Hunter and Priest 1960; Richardson et al. 1963).

These raw measurements were also edited using the Churchill program, and any values more than three standard deviations from their predicted values were re-measured. In the cases where errors were found, the necessary corrections were made.

More than half of the data records had missing values for the third molars. This was due mainly to the age distribution of our sample population, with a median age of less than sixteen years. Missing observations occurred in random fashion for most of the remaining teeth, affecting a maximum of 12–15 percent of the subjects on certain measurements. Mindful of the trade-off between complete data records and reduction in sample size, regression techniques using the entire Bougainville sample were used to estimate the size of the missing teeth, based on the size of two adjacent and highly correlated teeth. If, however, there were more than four tooth measurements missing in a particular individual, he or she was excluded from the sample. The data for males and females were analyzed separately not only because doing so provided a biological measure of repeatability for the village arrays, but also because there is considerable sexual dimorphism in the size of teeth and especially in the canines.

Dermatoglyphics
As mentioned in Chapter 6, two students, Christopher Hitt and Brian Eisenberg, performed the laborious classification and ridge counting of the palm- and finger-prints, under the supervision of Jeffrey Froehlich. Those below ten years of age were excluded from the analysis, mainly because we decided that our sample was large enough, and the task of analyzing the prints was so tedious that we had all reached the point of diminishing returns. Because the original sample sizes were so large, all individuals with imperfect prints were excluded from the analysis.

Although a more detailed classification of ridge types was actually performed, the categories of pattern types used in the final analysis were: arch, tented arch, radial loop, ulnar loop, composites (two or more triradii present), and true whorls. These classifications, described in Chapter 6, are explained in more detail in Cummins and Midlo (1961). On the palm, the classification of the mainlines A, B, C, D, and T was included. Ridge counts were made on ulnar and radial sides of each digit. This allowed the calculation of a "maximum" ridge count for each finger, meaning, according to Bonnevie's usage, the largest ridge count for each finger, and the "total" ridge count, here meant as the sum of both radial and ulnar ridge counts for each finger.

Blood Chemistries
The whole-blood samples were analyzed for red cell antigens and selected serum proteins by R. J. Walsh's group in Sydney, first at the Australian Red Cross, and then at the University of New South Wales.

The procedures followed for the ABO, MNSs, P, Kell, and Duffy typing at Walsh's laboratory are described in Giles et al. (1966). Also, the analyses of the acid phosphatase, haptoglobin, and transferrin systems were undertaken by Walsh's group, in which Dr. L. Y. C. Lai played a prominent role. P typing was carried out only on the first third of the total sample, and as the determinations were evidently more questionable than usual, their results have been discarded. Red cell acid phosphatase tests were completed on only 1165 of the 2009 samples, but otherwise all noncontaminated samples were typed for the complete battery. The appropriate references for these methods are: For PHs (Hopkinson et al. 1963, 1964); haptoglobin (Smithies 1955); and transferrin (Mueller et al. 1962).

After all these tests were completed, the remaining sera were packed in dry ice and flown to Dr. Arthur G. Steinberg's laboratory at the Department of Biology, Case Western Reserve University, Cleveland. Dr. Steinberg's group carried out the determinations of the Gm and Inv antigenic factors. All samples were tested for the Gm antigens (1, 2, 3, 4, 5, 13, and 14), and for Inv (1), all as previously described (Steinberg 1962).

Sera were also sent to Dr. Baruch Blumberg's laboratory at the Institute for Cancer Research in Philadelphia, where testing for the Australia antigen was performed (see Blumberg et al. 1971). As the genetic basis of this trait is still unclear, it again will not be discussed in this text.

The results of all these tests on each individual were sent to me in Cambridge, and I calculated both phenotype and allele frequencies for all systems and groups. For the calculation of all the allele frequencies and related parameters (chi-square "goodness of fit," standard deviations of allele estimates, and observed and expected numbers), I adapted the Fortran II MAXIM program written by Kurczynski and Steinberg (1967). This program gives the maximum likelihood estimate of allele frequencies and greatly reduces the possibility of error in the values reported.

These allele frequencies for each village were the data necessary for the calculation of village gene frequency variances and covariances around the total population means, and this in turn allowed for the construction of genetic distances and spatial representations of the village relationships, described below.

III. Statistical Analysis and Comparison

Biological Distances Using Proportions and Frequencies

There are currently a bewildering array of statistical approaches for describing and analyzing sets of data of this sort, and I readily admit to having taken two years to decide upon the methods which were most appropriate. As I was most interested in comparing the patterns of variation in the different data sets, I ultimately decided that it was most straightforward to generate, from each set, spatial (ordinarily multidimensional) arrays of the village samples according to their relative differences. These patterns of differences, or "biological distances," could then be compared with one another using the technique most commonly referred to as Gower's R^2.

This required obtaining Euclidean distances among the different groups. Referring to Figure A.1, the simplest sort of "distance" measure between two samples with means i and j should be,

$$d_{ij}^2 = a^2 + b^2. \tag{1}$$

This may be written in terms of the variances and covariances of the sample means from the origin, or population mean, as shown below.

By trigonometry,

$$d_{ij}^2 = a^2 + b^2.$$

But, also,

$$d_{ij}^2 = \text{var}_i + \text{var}_j - 2\,\text{cov}_{ij}.$$

For as

$$\text{var}_i = (a + c)^2 + d^2 \quad \text{[The squared deviation of } i \text{ from the origin]}$$

and

$$\text{var}_j = (b + d)^2 + c^2 \quad \text{[The squared deviation of } j \text{ from the origin]}$$

A.1 Geometric representation of finding a distance relationship between two points, i and j

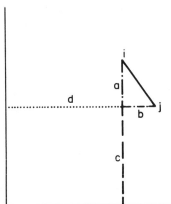

and

$$\text{cov}_{ij} = (d + b)d + (a + c)c,$$

then,

$$\text{var}_i + \text{var}_j - 2 \, \text{cov}_{ij}$$
$$= (a + c)^2 + d^2 + (b + d)^2 + c^2 - 2[d(d + b) + c(a + c)]$$
$$= a^2 + 2ac + c^2 + d^2 + b^2 + 2bd + d^2 + c^2 - 2d^2 - 2bd - 2ac - 2c^2$$
$$= a^2 + b^2.$$

For example, the **R** matrix calculated from intermarriage frequencies in Chapter 3 is essentially a predicted variance-covariance matrix (or matrix with variances in the diagonal and covariances between groups on the off-diagonals), and these can be converted into a symmetric matrix of distances between each pair of villages where

$$d_{ij}{}^2 = r_{ii} + r_{jj} - 2r_{ij}. \tag{2}$$

In this case, the resulting distances are a prediction of what the genetic relationships of the fifteen different villages would be if their current mating and migration patterns had been maintained over fifty generations, with no change in selection or systematic pressure and no change in village sample size, and if the current populations all had a homogeneous origin. Therefore, they do not really serve as a prediction of current genetic diversity at all; but they are interpreted as a "determinator" of diversity as are geographic measures of proximity, or linguistic proximity, for that matter.

For gene frequencies, as reported in Chapter 6, I prefer essentially the same statistic, equation (2), to calculate genetic distances, although in dealing with gene frequencies, the variances and covariances or sample differences must be standardized by some estimate of the population's mean gene frequency, because one property of multinomial sampling is that the magnitude of a difference in gene frequencies depends upon the initial, or population, frequency from which they now deviate.

The original genetic distance statistic of Sanghvi (1953) used the pairwise means of the frequencies of the two groups as the standardizing variables, so that the distance statistic took the form of a chi-square statistic. For, where

$$x^2 = \sum \frac{(\text{observed} - \text{expected})^2}{\text{expected}},$$

with summation over all categories or alleles, if "expected" is taken as the pairwise population mean and "observed" values are the two village sample frequencies, the equation becomes

$$X^2 = \frac{(p_i - \bar{p})^2}{\bar{p}} + \frac{(p_j - \bar{p})^2}{\bar{p}}.$$

If the overall mean is used in the denominator, then this becomes Sanghvi's G:

$$G = \sum \frac{(p_i - p_j)^2}{\bar{p}},$$

and is very closely related to the r statistic used in Chapter 4, which only differs in taking into consideration the sampling variances. There are a number of arguments over the appropriateness of such distance measures, but an extended discussion of these is not appropriate here. The interested reader is referred to Kurczynski (1970), Edwards and Cavalli-Sforza (1964), and Nei (1973) among others.

Metric Distances and Discriminant Analysis

The anthropometric, dentometric, and dermatoglyphic data sets were all subjected to a multiple discriminant analysis. Discriminant analysis is now a commonly used tool in physical anthropology, and excellent discussions are available in the literature (see especially Howells 1973). Originally developed to establish criteria for classifying test individuals into known groups, discriminant analysis here is used to describe the extent and the nature of the differences among the village samples. Differences among individuals within the total sample which do not serve to distinguish the different subsamples (for example, the age differences distinguished in the longitudinal anthropometric study in Chapter 5) are ignored, and, in the sort of analysis used here, it is assumed such variations are constant from one subgrouping to the next (or that the variances and covariances of the different samples are relatively homogeneous).

The mathematical procedure calls for separating the overall variances and co-variances of the measurements taken on all individuals into two matrices, the pooled within-group variances and covariances (W), and the among-group variances and covariances (A). The ratio of the among-group variation to the within-group variation, analogous to dividing A by W, is properly denoted in matrix notation by $W^{-1}A$. The resulting matrix can then be manipulated (or "factored") in any number of ways, but, most commonly, computation follows the principal components procedure. Successive major axes of variation (here read discrimination) are extracted from the matrix by solving the following equation:

$$(\mathbf{W}^{-1}\mathbf{A} - \lambda\mathbf{I})\mathbf{b} = 0$$

where $\mathbf{I}$ is an identity matrix, λ the latent roots to be found, and $\mathbf{b}$ the associated latent vector with each latent root. There is produced a complete set of latent roots of descending size, equal in number to one less than the number of groups or measurements. These are properly a diagonal matrix $\mathbf{L}$ with the roots as the diagonal elements. The latent vectors properly appear as columns in an associated matrix $\mathbf{Q}$.

The latent vectors act as direction cosines which rotate the original axes (measurements) of variation to new positions. The new axes, which should be almost perfectly orthogonal, or at right angles to one another, are hierarchical in nature, the first being the most important, or accounting for the greatest amount of variation. The latent roots indicate the relative importance of the different new axes.

Converting an individual's vector of measurements in the original test space to scores or positions in the newly rotated discriminant space is achieved by substituting the measurements as follows:

$$y_{1i} = X_{j1i}b_{j11} + X_{j2i}b_{j21} \ldots + X_{jpi}b_{jp1},$$

where y_{1i} is the discriminant score for individual i on the first discriminant func-

A.2 The effect of rotating two congruent triangles in a plane. On the left, the two will fit exactly. On the right, the fit is poor unless rotation in three dimensions is allowed, in which case the fit is again exact

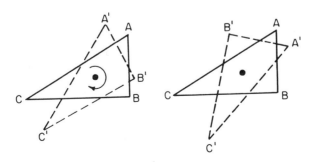

tion; b is the particular coefficient for the measurement, j, out of p measurements for the first discriminant; and $X_1 \ldots _p$ are the original raw measurements taken on individual i.

Since the coefficients are multiplied by the raw scores, they are necessarily to the scale of the different measurements, and are most difficult to interpret. Scaled vectors are sometimes calculated to remove the differences in scale (using the pooled within-group estimate of that measurement's variance), but they are also relatively hard to interpret biologically. The most easily understandable indication of what measurements a discriminant function is related to is given by "discriminant loadings," which are simply the correlations of raw measurements with discriminant function scores. These are used in the text.

Because the discriminant functions will account for the differences between village or sample mean vectors in a hierarchical fashion (meaning that the first function will account for the maximum village differences, the second function the greatest part of the remaining differences, and so on), they give the best representation of (Mahalanobis') D^2 relationships in restricted dimensionality. When the discriminant functions are appropriately scaled, the village or sample means on the functions are separated by Mahalanobis' D^2 (see, for example, Gower 1966).

Generating a Multidimensional Representation of a Set of Distances by Principal Coordinates Analysis

If a matrix D of pairwise distances between all pairs of populations is given, as in the case of the matrix of unshared cognate percentages utilized in Chapter 2, where the villages can be represented by points P_j or P_k, and j and k take the values 1, 2 . . . n; then D is a symmetric matrix of order n with zeroes down the main diagonal. To obtain coordinates of every population with principal axes as orthogonal coordinate axes, a series of transformations have to be made (Gower 1966a).

(1) Define a matrix E with elements $-\frac{1}{2} d_{jk}^2$;

(2) Writing $e_{.k}, e_{j.}$ and $e_{..}$ for the row, column, and general means of E, evaluate a new normalized matrix F whose elements f_{jk} are

$$e_{jk} - e_{j.} - e_{.k} + e_{..}$$

(3) Find the latent roots and vectors (λ and X) of F. In matrix notation,

$$FX = XL,$$

L is a diagonal matrix of the latent roots with the diagonals ($l_1, l_2 \ldots , l_n$) assumed to be in decreasing size, so that l_i is the sum of squares about the mean of the coordinates in the ith dimension and therefore $l_n = 0$ because n points can always be represented in $n-1$ dimensions.

(4) Scale the columns of $\mathbf{X}$, so that the sum of squares of the ith column is l_i, the ith largest latent root. Thus,

$$\mathbf{X'X} = \mathbf{L} \text{ and } \mathbf{XX'} = \mathbf{F}.$$

Then the elements of the ith row of $\mathbf{X}$ are the required coordinates of P_i.

Comparison of Village Sample Arrays in Different Test Spaces

The method I have chosen to use is by no means perfect, as the discussion in Chapter 8 makes clear. However, I think it has definite advantages over other approaches I am acquainted with. The method I have chosen, espoused recently by Gower (1970) rotates the configuration of points derived from one analysis until they best "fit" the solution of another analysis, possibly allowing for a magnification factor at the same time. Figure A.2 illustrates the problem of "matching" two identical triangles, or the coordinates of their angles, when one of the triangles must be rotated and also turned over, or "perverted," to achieve the best fit. According to this technique, which allows for standardization of scales between figures or configurations, congruent triangles or figures should be perfectly matched, no matter what the original scales were. This is not necessarily a desirable feature in all cases, but it seems most appropriate in the present situation, when we are comparing geographic distances with anthropometric, linguistic, and blood genetic distances.

After rotation, the sum of squares of the distances between pairs of like points (in the current instance the distance between the coordinates representing village 1 in both configurations, village 2 in both configurations, and so forth) gives a statistic (R^2) which measures the goodness of fit of the configurations. This method has been used by Sneath (1967) to compare homologous points on pairs of skulls in a two-dimensional representation. The generalization to more than two dimensions of this procedure is given by Gower (1970) without proof, and details of the method are given in Gower (1971).

In outline in matrix notation, the procedure is as follows. The first, or reference, configuration has n populations or units represented as coordinates in a space of p dimensions; and the second, or alternative, configuration, represents the same n units in another geometric space of q dimensions. The coordinates of the points in the two multidimensional spaces constitute two matrices, $\mathbf{X}$ and $\mathbf{Y}$, $\mathbf{X}$ being the first configuration of points, the $n \times p$ matrix, and the ith row being the coordinates of the ith population in this first configuration. $\mathbf{Y}$ is the $n \times q$ matrix with each row the coordinates of a population in the second coordinate system. All coordinates are assumed to be referred to orthogonal axes with zero means, as is the case with results from principal coordinates analysis.

The rotation of $\mathbf{Y}$ that will give the optimal fit in a least-squares sense to $\mathbf{X}$ can be represented by an orthogonal matrix $\mathbf{H} = \mathbf{VU}'$, where $\mathbf{V}$ and $\mathbf{U}$ are obtained from the singular value decomposition of $\mathbf{X'Y}$. The singular value decomposition of $\mathbf{X'Y}$ is $\mathbf{X'Y} = \mathbf{U\Sigma V}'$, where $\mathbf{U}$ and $\mathbf{V}$ are orthogonal matrices, and $\mathbf{\Sigma}$ is a diagonal matrix of the singular values of $\mathbf{X'Y}$.[3] The reflections required to take care of the arbitrary signs of $\mathbf{V}$ and $\mathbf{U}$ are introduced by means of a diagonal matrix with diagonal elements either $+1$ or -1, the sign depending on the sign of the corresponding elements of $\mathbf{S} = \mathbf{U'X'YV}$. $\mathbf{H}$ is then computed $\mathbf{H} = \mathbf{VSU}'$.

The statistical properties of the goodness of fit statistic $R^2 = $ trace $(\mathbf{XX}' + \mathbf{YY}' - 2\mathbf{YHX}')$ are essentially unexplored. Still it avoids the clearly untenable assumptions of the correlation method.

3. For computational methods, see Golub, 1968; Golub and Reinsch, 1970.

Bibliography

Allen, J. and C. Hurd
n.d. *Languages of the Bougainville District.* Summer Institute of Linguistics, Ukarumpa, New Guinea.

Armor, D. J., and A. Couch
1971. *The Data-Text Primer.* New York, Free Press.

Bailit, H. L., and B. Sung
1968. "Maternal effects on the dentition," *Archives of Oral Biology, 13:*155–162.

Balakrishnan, V., and L. Sanghvi
1968. "Distance between populations on the basis of attribute data," *Biometrics, 24:*857–865.

Blackwood, B.
1931. "Mountain people of the South Seas," *Natural History, 31:*424–433.
1935. *Both sides of Buka Passage: An ethnographic study of social, sexual, and economic questions in the northwestern Solomon Islands.* Oxford, Clarendon Press.

Blumberg, B. S., J. S. Friedlaender, A. Woodside, A. I. Sutnick, and W. T. London
1969. "Hepatitis and Australia Antigen: Autosomal recessive inheritance of susceptibility to infection in humans," *Proceedings of the National Academy of Sciences, 62:*1108–1115.

Bodmer, W., and L. L. Cavalli-Sforza
1968. "A migration matrix model for the study of random genetic drift," *Genetics, 59:*565–592.

Booth, P. B., and A. P. Vines
1967. "Blood groups and other genetic data from Bougainville, New Guinea, with observations on the occurrence of the R_o (cDe) and R_z (CDE) gene complexes in Melanesia," *Archaeology and Physical Anthropology in Oceania, 2:*227–235.

Capell, A.
1962. "Oceanic linguistics today," *Current Anthropology, 3:*371–428.

Cavalli-Sforza, L. L.
1966. "Population structure and human evolution," *Proceedings of the Royal Society of London,* Ser. B, *164:*362–379.
1969. "Human diversity," *Proceedings of the XIIth International Congress of Genetics, 3:*405–416.
1973. "Analytic review: Some current problems of human population genetics," *American Journal of Human Genetics, 25:*82–104.

Cavalli-Sforza, L. L., and W. Bodmer
1971.   *The Genetics of Human Populations.* San Francisco, Cal., Freeman.
Cavalli-Sforza, L. L., L. A. Zonta, F. Nuzzo, L. Bernini, W. W. W. DeJong, P. Meera Kahn, A. K. Ray, L. N. Went, M. Siniscalco, L. E. Nijenhuis, E. van Loghem, and A. Modiano
1969.   "Studies on African Pygmies. I: A pilot investigation of Babinga Pygmies in the Central African Republic (with an analysis of genetic distances)," *American Journal of Human Genetics, 21:*252–274.
Chai, C. K.
1967.   *Taiwan Aborigines.* Cambridge, Mass., Harvard University Press.
Chinnery, E. W. P.
1924.   "The natives of south Bougainville and Mortlocks (Taku)," *Territory of New Guinea Anthropological Report,* No. 5, Canberra, pp. 87–108.
Churchill, E.
1963.   "The modern computer and physical anthropology," *American Journal of Physical Anthropology, 21:*426.
Conneally, P. M., A. D. Merritt, B. E. Quinn, and R. H. Y. Potter
1968.   "Semiautomatic digital printing caliper for tooth measurements," *Journal of Dental Research, 47:*501.
Cooley, W. W., and P. R. Lohnes
1962.   *Multivariate Procedures for the Behavioral Sciences.* New York, John Wiley.
Cowgill, U. A.
1966.   "The season of birth in man," *Man,* n. s., *1:*232–240.
Cummins, H., and C. Midlo
1961.   *Finger Prints, Palms and Soles.* New York, Dover.
Damon, A.
1974.   "Human ecology in the Solomon Islands. Biomedical observations among four tribal societies," *Journal of Human Ecology, 2:*191–215.
Dietz, E. J., H. L. Bailit, and D. Kolakowski
1974.   "A comparison of methods for estimating heritability, utilizing dental variables." Paper presented at the American Association of Physical Anthropologists' Annual Meeting, Amherst, Massachusetts.
Edwards, A. W. F., and L. L. Cavalli-Sforza
1964.   "Reconstruction of phylogenetic trees." In V. H. Heywood and J. McNeill, eds., *Phenetic and Phylogenetic Classification.* Systematics Association, Publication No. 6:*67–76. London.

Emmanuel, I., and J. Biddulph
1967.   "Child health among the Nasioi and Kwaio of the Solomon Islands." Ms.
Fisher, R. A.
1930.   *Genetical Theory of Natural Selection*. Oxford, Clarendon Press.
Fitch, W. M., and E. Margoliash
1967.   "Construction of phylogenetic trees," *Science, 155:*279–284.
Fortes, M.
1949.   *The Web of Kinship Among the Tallensi*. London and New York, Oxford University Press.
Friedlaender, J. S.
1969.   *Biological Diversity over Population Boundaries in south-central Bougainville*. Unpublished PhD thesis, Harvard University, Cambridge, Mass.
1971a.   "The population structure of south-central Bougainville," *American Journal of Physical Anthropology, 35:*13–26.
1971b.   "Isolation by distance in Bougainville," *Proceedings of the National Academy of Science, 68:*704–707.
1974.   "A comparison of population distance and kinship bioassay techniques." In J. F. Crow and C. Dennison, eds., *Population Distance*, New York, Plenum.
Friedlaender, J. S., and D. L. Oliver
1975.   "Effects of aging and secular increase in Bougainville men." In E. Giles and J. S. Friedlaender, eds., *Measures of Man*, Cambridge, Mass., Schenkman. (In press)
Friedlaender, J. S., L. A. Sgaramella-Zonta, K. K. Kidd, L. Y. C. Lai, P. Clark, and R. J. Walsh
1971.   "Biological divergences in south-central Bougainville: an analysis of blood polymorphism, gene frequencies and anthropometric measurements utilizing tree models, and a comparison of these variables with linguistic, geographic, and migrational distances," *American Journal of Human Genetics, 23:*253–270.
Frizzi, E.
1913.   "Reiseeindrücke aus Buka und Bougainville," *Mitteilungen der Geographischen Gesellschaft in München, 8:*484–490.
1914.   "Ein Beitrag zur Ethnologie von Bougainville und Buka, mit Spezieller Berücksichtigung der Nasioi," *Baessler-Archiv*, VI, Leipzig and Berlin, B. G. Teubner.

Froehlich, J.
1975.   "The inheritance of continuous human traits: the quantitative genetic study of fingerprints." In E. Giles and J. S. Friedlaender, eds., *Measures of Man,* Cambridge, Mass., Schenkman. (In press)
Gallango, M. L., and T. Arends
1967.   "Inv(a) serum factor in Venezuelan Indians," *Vox Sanguinis, 13:* 457–460.
Garn, Stanley M., A. B. Lewis and R. S. Kerewsky
1968.   "Relationship between the buccolingual and mesiodistal tooth diameters," *Journal of Dental Research, 47:495.*
Giblett, E. R.
1969.   *Genetic Markers in Human Blood.* Oxford and Edinburgh, Blackwell Publications.
Giles, E., E. Ogan, R. J. Walsh, and M. A. Bradley
1966.   "Blood group genetics of natives of the Morobe District and Bougainville, Territory of New Guinea," *Archaeology and Physical Anthropology in Oceania, 1:*135–154.
Golub, G. H.
1968.   "Least squares, singular values and matrix approximations," *Aplikace Matematiky, 13:*44–51.
Golub, G. H., and C. Reinsch
1970.   "Handbook series linear algebra. Singular value decomposition and least squares solutions," *Numerische Mathematik, 14:*403–420.
Goose, D. H.
1967.   "Preliminary studies of tooth size in families," *Journal of Dental Research, 46:*959–962.
Gower, J. C.
1966.   "Some distance properties of latent roots and vector methods used in multivariate analysis," *Biometrika, 53:*325–338.
1970.   "Classification and geology," *International Statistics Institute Review, 38:*35–41.
1971.   "Statistical methods of comparing different multivariate analyses of the same data." In F. R. Hodson, D. G. Kendall, and P. Tautu, eds., *Mathematics in the Archaeological and Historical Sciences,* Edinburgh, Edinburgh University Press, pp. 138–149.
Guppy, H. B.
1887.   *The Solomon Islands and Their Natives.* London, Swan, Sonnenschein.
Hamilton, L., and W. Wilson
1957.   "Dietary survey in Malaguna village, Rabaul," *Food and Nutrition Notes*

and Review, Commonwealth Department of Health. Issued by Australia Institute of Anatomy, Canberra, *14:*77.

Harpending, H., and T. Jenkins
1973.   "Genetic distance among southern African populations." In M. H. Crawford and P. L. Workman, eds., *Method and Theories of Anthropological Genetics,* University of New Mexico Press, pp. 177–200.
1974.   "Kung Population Structure." In J. F. Crow and C. Dennison, eds., *Genetic Distance.* New York, Plenum.

Harpending, H., and W. Chasco
1975.   "Heterozygosity and population structure in Southern Africa." In E. Giles and J. S. Friedlaender, eds., *Measures of Man,* Cambridge, Mass., Schenkman. (In press)

Harris, D.
1969.   "NUMIX." In N. E. Morton, ed., *A Genetics Program Library,* Honolulu, University of Hawaii Press, p. 41.

Harris, H., and D. A. Hopkinson
1972.   "Average heterozygosity per locus in man: an estimate based on the incidence of enzyme polymorphisms," *Annals of Human Genetics (London), 36:*9–20.

Hiernaux, J.
1956.   "Analyse de la variation des caractères physiques humains en une région de l'Afrique centrale: Ruanda-Urundi et Kivu." In *Annales du Musée royal du Congo Belge. Serie en 8ᵉ Sciences de l'homme,* vol. 3. Turvuren.
1963.   "Heredity and environment: their influences on human morphology. A comparison of two independent lines of study," *American Journal of Physical Anthropology, 21:* 575–589.

Holt, S.
1968.   *The Genetics of Dermal Ridges.* Springfield, Ill., Thomas.

Hopkinson, D. A., N. Spencer, and H. Harris
1963.   "Red cell acid phosphatase variants; a new human polymorphism," *Nature* (London) *199:* 969.
1964.   "Genetical studies on human red cell acid phosphatase," *American Journal of Human Genetics, 16:*141.

Howell, N.
1973.   "The feasibility of demographic studies, in 'anthropological' populations." In M. H. Crawford and P. L. Workman, eds., *Method and Theories of Anthropological Genetics,* University of New Mexico Press, pp. 249–262.

Howells, W. W.
1951.   "Factors of human physique," *American Journal of Physical Anthropology,* 9:159–192.
1966a.   "Population distances: biological, linguistic, geographical, and environmental," *Current Anthropology, 7:*531–540.
1966b.   "Variability in family lines vs. population variability," *New York Academy of Sciences Annals, 134:*624–631.
1973.   *Cranial Variation in Man.* Peabody Museum Papers, Harvard University, No. 67. Cambridge, Mass.
Hunt, E. E., and J. D. Mavalwala
1964.   "Finger ridge counts in the Micronesians of Yap," *Micronesia, 1:*55–58.
Hunter, W. S., and W. R. Priest
1960.   "Errors and discrepancies in measurement of tooth size," *Journal of Dental Research, 39:*405–414.
Imaizumi, Y., N. E. Morton, and D. E. Harris
1971.   "Isolation by distance in artificial populations," *Genetics, 66:*569–582.
Johnson, G. B.
1974.   "Enzyme polymorphism and metabolism," *Science, 184:*28–37.
Kalmus, H.
1958.   "Improvements in the classification of the taster genotypes," *Annals of Human Genetics, 22:*222–230.
Kariks, J. O. Kooptzoff, and R. J. Walsh
1957.   "Blood groups of the native inhabitants of Bougainville, New Guinea," *Oceania, 28:*146–158.
King, J. L., and T. H. Jukes
1969.   "Non-Darwinian evolution," *Science, 164:*788–798.
Kirk, R. L.
1965.   "Population genetic studies of the indigenous peoples of Australia and New Guinea." In A. G. Steinberg and A. Bearn, eds., *Progress in Medical Genetics, 4,* New York, Grune and Stratton, pp. 202–241.
Kirk, R. L. and N. M. Blake
1971.   "Population genetic studies in Australian aborigines of the Northern Territory. The distribution of some serum protein and enzyme groups among populations at various localities in the Northern Territory of Australia," *Human Biology in Oceania, 1:*54–76.
Kohne, D. E.
1970.   "Evolution of higher-organism DNA," *Quarterly Review of Biophysics, 43:*327–375.

Kraus, B. S., and R. E. Jordan
1965.   *The Human Dentition Before Birth,* Philadelphia, Pa., Lea and Febiger.
Kurczynski, T., and A. G. Steinberg
1967.   "A general program for maximum likelihood estimation of gene frequencies," *American Journal of Human Genetics, 19:*178–179.
Langaney, A., J. Gomila, and C. J. Bouloux
1972.   "Bedik: Bioassay of kinship," *Human Biology, 44:*475–488.
Langaney, A., and J. Gomila
1973.   "Bedik and Niokholonko intra- and inter-ethnic migration," *Human
Biology, 45:*137–150.
Lerner, I. M.
1954.   *Genetic Homeostasis.* Edinburgh, Oliver and Boyd.
1969.   *Heredity, Evolution, and Society.* San Francisco, Cal., Freeman.
Lewontin, R. C.
1967.   "An estimate of average heterozygosity in man," *American Journal of
Human Genetics, 19:*681–685.
1972.   "The apportionment of human diversity," *Evolutionary Biology, 6:*381–
398.
Lewontin, R. C., and J. Krakauer
1973.   "Distribution of gene frequency as a test of the theory of selective neu-
trality of polymorphisms," *Genetics, 74:*175–195.
Littlewood, R.
1973.   *Physical Anthropology of New Guinea Highlands.* Seattle, University of
Washington Press.
Loesch, D.
1971.   "Genetics of dermatoglyphic patterns on palms," *Annals of Human
Genetics, 34:*277–293.
Long, G.
1963.   *The Final Campaigns.* Canberra, Australian War Memorial.
MacLean, C.
1969.   "Computer analysis of pedigree data." In N. E. Morton, ed., *Computer
Applications in Genetics,* Honolulu, University of Hawaii Press, pp. 82–86.
Main, J. H. P.
1966.   "Retention of potential to differentiate in long-term cultures of tooth
germs," *Science, 152:*778–779.
Mair, L.
1948.   *Australia in New Guinea.* London, Christophers.
Majumdar, D. N., and C. R. Rao

1960.   *Race Elements in Bengal: A Quantitative Study.* Calcutta.

Malécot, G.
1948.   *Les Mathématiques de l'hérédité.* Paris, Masson et Cie.

Mayr, E.
1963.   *Animal Species and Evolution.* Cambridge, Mass., Harvard University Press.

McHenry, H., and E. Giles
1971.   "Morphological variation and heritability in three Melanesian populations: a multivariate approach," *American Journal of Physical Anthropology, 35:*241–253.

Mitchell, D. D.
1972.   *Gardening for Money: Land and Agriculture in Nagovisi.* Unpublished PhD thesis, Harvard University, Cambridge, Mass.

Moorrees, C. F., E. A. Fanning, and C. H. Hunt
1963.   "Formation and resorption of three deciduous teeth in children," *American Journal of Physical Anthropology, 21:*205–213.

Morton, N. E.
1968.   "Bioassay of population structure under isolation by distance," *American Journal of Human Genetics, 20:*411–419.
1969.   "Human population structure," *Annual Review of Genetics, 3:*53–74.

Morton, N. E., E. Krieger, and M. P. Mi
1966.   "Natural selection on polymorphisms in northeastern Brazil," *American Journal of Human Genetics, 18:*153–171.

Morton, N. E., and J. M. Lalouel
1972.   "Topology of kinship in Micronesia," *American Journal of Human Genetics, 25:*422–432.

Mueller, J. O., O. Smithies, and M. R. Irwin
1962.   "Transferrin variation in Columbidae," *Genetics, 47:*1385–1392.

Nash, J.
1972.   *Aspects of Matriliny in Nagovisi Society.* Unpublished PhD Thesis, Harvard University, Cambridge, Mass.

Neel, J. V.
1970.   "Lessons from a 'primitive' people," *Science, 170:*815–822.

Neel, J. V., F. M. Salzano, P. C. Junqueira, F. Keiter, and D. Maybury-Lewis
1964.   "Studies on the Xavante Indians of the Brazilian Mato Grosso," *American Journal of Human Genetics, 16:*52–140.

Neel, J. V., and M. Salzano
1966.   "Further studies on the Xavante Indians. X: Some hypotheses—generalizations resulting from these studies," *American Journal of Human Genetics, 19:554–574.*

Neel, J. V., and W. J. Schull
1972.   "Differential fertility and human evolution," *Evolutionary Biology, 6:363–380.*

Neel, J. V., and R. J. Ward
1972.   "The genetic structure of a tribal population, the Yanomama Indians. VI: Analysis by *F*–statistics (including a comparison with the Makiritare and Xavante," *Genetics, 72:639–666.*

Nei, M., and A. K. Roychoudhury
1974.   "Sampling variances of heterozygosity and genetic distances," *Genetics, 76:379–390.*

Ogan, E.
1972.   *Business and Cargo: Socio-economic Change among the Nasioi of Bougainville,* New Guinea Research Bulletin No. 44, Australian National University, Port Moresby and Canberra.

Ogan, E., J. Nash, and D. D. Mitchell
1975.   "Culture change and fertility in two Bougainville populations." In E. Giles and J. S. Friedlaender, eds., *Measures of Man,* Cambridge, Mass., Schenkman.

Oliver, D. L.
1949.   *Studies in the Anthropology of Bougainville, Solomon Islands,* Peabody Museum Papers, vol. 29, nos. 1–4, Harvard University, Cambridge, Mass.
1954.   *A Solomon Island Society: Kinship and Leadership among the Siuai of Bougainville.* Cambridge, Mass., Harvard University Press.
1955.   *Somatic Variability and Human Ecology on Bougainville, Solomon Islands.* Peabody Museum, Cambridge, Mass. Ms.
1973.   *Bougainville: A Personal History.* Melbourne, Melbourne University Press.

Oliver, D. L., and W. W. Howells
1957.   "Micro-evolution: cultural elements in physical variation," *American Anthropologist, 59:965–978.*
1960.   "Bougainville populations studied by generalized distance," In *Actes du VIᵉ Congres International des Sciences Anthropologiques et Ethnologiques,* vol. I, Paris.

Osborne, R. H., and F. V. DeGeorge
1959.   *Genetic Basis of Morphological Variation.* Cambridge, Mass., Harvard University Press.

Osborne, R. H., S. L. Horowitz, and F. V. DeGeorge
1958.   "Genetic variation in tooth dimensions: A twin study of the permanent anterior teeth," *American Journal of Human Genetics, 10:*350–356.

Page, L., A. Damon, and W. Moellering
1974.   "Antecedents of cardiovascular disease in six Solomon Islands societies," *Circulation,* June 1974.

Parkinson, R.
1907.   *Dreissig Jahre in der Südsee: Land und Leute, Sitten und Gebraüche im Bismarckarchipel und auf den deutschen Salomoinseln.* Stuttgart, Strecher und Schroder.

Parsi, P., and M. DiBacco
1968.   "Fingerprints and the diagnosis of zygosity in twins," *Acta Genetica Medica Gemellae, 17:*333–358.

Parsons, P. A.
1964.   "Finger-print pattern variability," *Acta Genetica et Statistica Medica Basel, 14:*201–211.

Penrose, L. S.
1954.   "Distance, size and shape," *Annals of Eugenics, 18:*337–343.

Pollitzer, W. S.
1958.   "The Negroes of Charleston [S.C.]: a study of hemoglobin types, serology, and morphology," *American Journal of Physical Anthropology, 16:*213–234.

Pons, J.
1959.   "Quantitative genetics of palmar dermatoglyphics," *American Journal of Human Genetics, 11:*252–256.

Potter, R. H. Y., P. L. Yu, A. A. Dahlberg, A. D. Merritt, and P. M. Conneally
1968.   "Genetic studies of tooth size factors in Pima Indian families," *American Journal of Human Genetics, 20:*89–100.

Post, R. H.
1962.   "Population differences in red and green color vision deficiency: a review, and a query on selection relaxation," *Eugenics Quarterly, 9:*131–146.

Prakash, S.
1973.   "Patterns of gene variation in central and marginal populations of

Drosophila robusta," *Genetics, 75:*347–369.

Rappaport, R.
1967.   *Pigs for Ancestors: Ritual in the Ecology of a New Guinea People.* New Haven, Conn., Yale University Press.

Rhoads, J. G.
1972.   *Factors of physique in adult males in the United States and the Solomon Islands.* Unpublished B.A. thesis, Peabody Museum, Harvard University, Cambridge, Mass.

Rao, C. R.
1963.   *Advanced Statistical Methods in Biometric Research* (rev. ed.), New York, John Wiley.

Ribbe, C.
1903.   *Zwei Jahre Unter den Kannibalen der Salomo-Inseln. Reiseerlebnisse und Schilderungen von Land und Leuten.* Dresden-Blasewitz, Hermann Beyer.

Richardson, M. E., C. P. Adams, and T. P. G. McCartney
1963.   "Analysis of tooth measuring methods on dental casts," European Orthodontic Society.

Rohlf, F. J., and G. D. Schnell
1971.   "An investigation of the isolation by distance model," *American Naturalist, 105:*295–324.

Schlaginhaufen, O.
1908.   "Bericht über eine Orientierungsreise nach Kieta auf Bougainville," *Zeitschrift für Ethnologie, 40:*85–86.

Scott, R. M., P. B. Heyligers, J. R. McAlpine, J. C. Saunders, and J. G. Speight
1967.   *Lands of Bougainville and Buka Islands, Territory of New Guinea.* Land Research Series No. 20, Commonwealth Scientific and Industrial Research Organisation, Melbourne.

Scragg, R. F. R.
1954.   *Depopulation in New Ireland. A Study of Demography and Fertility.* Government Printer, Port Moresby. As cited in Vines 1970.

1967.   Unpublished tables of vital events in the New Ireland and Bougainville area. As cited in Vines 1970.

Simmons, R. T., V. Zigas, L. L. Baker, and D. C. Gajdusek
1961.   "Studies on Kuru. V. A blood group genetical survey of the Kuru region and other parts of Papua-New Guinea," *American Journal of Tropical Medicine and Hygiene, 10:*639–664.

Slatkin, M.
1970.   "Selection and polygenic characters," *Proceedings of the National Academy of Science, 66:87–93.*
Smithies, O.
1955.   "Zone electrophoresis in starch gels: group variations in the serum proteins of normal human adults," *Biochemical Journal, 61:629–641.*
Sneath, P. A.
1967.   "Trend surface analysis of transformation grids," *Journal of Zoological Society* (London), *151:65–122.*
Sofaer, J. A., J. D. Niswander, C. J. MacLean, and P. L. Workman
1972.   "Population studies of south-western Indian tribes. V. Tooth Morphology as an indicator of biological distance," *American Journal of Physical Anthropology, 37:357–366.*
Soulé, M.
1971.   "The variation problem: the gene flow-variation hypothesis," *Taxon, 20:* 37–50.
Steinberg, A. G.
1962.   "Progress in the study of genetically determined human gamma globulin types (the Gm and Inv groups). In A. G. Steinberg and A. G. Bearn, eds., *Progress in Medical Genetics, 2,* New York, Grune and Stratton, pp. 1–38.
Stern, C.
1960.   *Principles of Human Genetics.* 2nd ed., San Francisco, Cal., Freeman.
Swadesh, M.
1955.   *Amerindian Non-Cultural Vocabularies.* rev. ed., Denver, Colo.
Tenczar, P., and R. S. Bader
1966.   "Maternal effects in dental traits of the house mouse," *Science, 152:1398–* 1400.
Terrell, J.
1972.   "Reports of the Bougainville Archaeological Survey, Number 5: Geographic systems and human diversity in the northern Solomons." Field Museum of Natural History, Chicago (Mimeo). Read at A. A. A. meetings, 1972.
Terrell, J., and G. J. Irwin
1972.   "History and tradition in the northern Solomons: an analytical study of the Torau migration to southern Bougainville in the 1860's," *The Journal of the Polynesian Society, 81:317–349.*

Thurnwald, R.
1912.   *Forschungen auf den Salomoinseln und dem Bismarck-Archipel*, 2 vols., Berlin, Dietrich Reimer Verlag.

Van Valen, L.
1965.   "Morphological variation and width of ecological niche," *The American Naturalist, 94:*377–390.

Vines, A. P.
1970.   *An Epidemiological Sample Survey of the Highlands, Mainland, and Island Regions of the Territory of Papua and New Guinea*, Port Moresby, Bloink.

Wahlund, S.
1928.   "Zusammensetzung von Populationen und Korrelationserscheinungen vom Standpunkt der Vererbungslehre aus betrachtet," *Hereditas, 2:*65–106.

Ward, R. H.
1972.   "The genetic structure of a tribal population, the Yanomama Indians. V:Comparison of a series of genetic networks," *Annals of Human Genetics, 36:*21–43.

Ward, R. H., and J. V. Neel
1970.   "Gene frequencies and micro-differentiation among the Makiritare Indians. IV:A comparison of a genetic network with ethnohistory and migration matrices; a new index of genetic isolation," *American Journal of Human Genetics, 22:*538–561.

Weil, G. J.
1971.   *Biological Consequences of Acculturation: A Study of the Eskimos of Northern Labrador*. Unpublished B. A. thesis, Peabody Museum, Harvard University, Cambridge, Mass.

Weninger, M.
1964.   "Zur 'polygenen' (additiven) Vererbung des Quantitativen Wertes der Fingerbeerenmuster," *Homo, 15:*96–103.

Wheeler, G. C.
1908.   *Mono-Alu Folklore (Bougainville Strait, Western Solomons)*, London, Routledge.

White, O.
1965.   *Parliament of a Thousand Tribes: A Study of New Guinea*. London, Heinemann.

Workman, P. L., B. S. Blumberg, and A. J. Cooper

1963.   "Selection, gene migration and polymorphic stability in a U. S. white and negro population," *American Journal of Human Genetics, 15:*429–437.

Workman, P. L., H. Harpending, J. M. Lalouel, C. Lynch, J. D. Niswander, and R. Singleton

1973.   "Population studies on southwestern Indian tribes. VI:Papago population structure: A comparison of genetic and migration analysis." In N. E. Morton, ed., *Genetic Structure of Populations.* Honolulu, University of Hawaii Press, pp. 166–194.

Workman, P. L., and J. D. Niswander

1970.   "Population studies on southwestern Indian tribes: II:Local genetic differentiation in the Papago," *American Journal of Human Genetics, 22:* 24–49.

Wright, S.

1943.   "Isolation by distance," *Genetics, 28:*114–138.

1948.   "On the roles of directed and random changes in gene frequency in the genetics of populations," *Evolution, 2:*279–294.

1949.   "Population structure in evolution," *Proceedings of the American Philosophical Society, 93:*471–478.

1951.   "The genetical structure of populations," *Annals of Eugenics, 15:*323–354.

1969.   "Evolution and genetics of populations." In *The Theory of Gene Frequencies,* vol. II. Chicago, Ill., University of Chicago Press.

Wurm, S.

1960.   "The changing linguistic picture of New Guinea," *Oceania, 31:*121–136.

Index